How We Built Our House for Under $10,000

Stephen T. Whitney

The "house on Nobody Hill" which the Walter Rae family built with their own hands in the township of Middlesex, Vermont. The Worcester Range looms in the background.

How We Built Our House for Under $10,000

by Walter Rae

A True-Life Adventure

Published MCMLXXIV by
YANKEE, INC.
Dublin, New Hampshire 03444

This book has been prepared
by the staff of
YANKEE, INC.
Editor: Clarissa M. Silitch
Art Director: Carl F. Kirkpatrick
Drawings by Carl F. Kirkpatrick

FIRST EDITION

Manufactured in the United States of America
Library of Congress Catalog Card No. 74-78890
ISBN No. 0-911658-61-0

For Steve and Walt Jr.,
my accomplices

Contents

Foreword

In every bookstore in the land, in libraries large or small, you'll find at least one book on how to build a house. But you'll never find one like HOW WE BUILT OUR HOUSE FOR UNDER $10,000.

You see, the other books are all written by professionals, and this book is written by an amateur. In fact, we're not sure Mr. Rae even qualified for that title at the outset of this project. By his own admission, his construction experience was limited to helping a friend build a woodland cabin—a simple shell without insulation.

So this is a book written by a man who started out as a rank beginner and became an authority on the experience of designing and building—from start to finish, through thick and thin (and thick again)—his own special house.

We repeat, there is not another book exactly like it in the country. And if you're one of the millions of people who have been imprisoned by specialization, you'll be on his side from page 10 to page 108.

Moreover, you'll learn a thing or two if you're in the planning stage of a similar project. You'll get a clear picture of the illimitable problems of constructing even a simple house (if there is such a thing). You'll discover that the advice of professionals is probably right and that experimentation is probably a gamble. You'll find out where you might waste time and money or both simultaneously, and finally, you may be able to determine more objectively whether or not you're the right man for such an undertaking.

In any case, you won't be bored. It reads like an adventure story—which is exactly what it is. It's an inside-out look at a project that touches on the very spirit of building and how that spirit affects and infects everyone involved.

If this book is singular, it is precisely because it *was* written by an amateur and because only the amateur sees everything. Architects, carpenters, plumbers, electricians are experts in their fields—but they're specialized and their view is limited. The carpenter usually doesn't concern himself with a design. He's more apt to resist it on the grounds that it's impractical (and quite often it *is*). And plumbers and electricians are historically indifferent to the overall aspects of any project they're working on.

Consider it—Mr. Rae chose his land, used his money, designed his house and built it . . . with his own two blistered hands. It was his dream that created the problem in the first place and his vision that held the whole hopelessly complicated mess together until the job was finished, and he could stand back and say, "Well done."

Now that's what we call an adventure story . . . in the great old Yankee hunch and by gorry tradition.

Austin N. Stevens
Managing Editor, Yankee *Magazine*
Dublin, New Hampshire

Introduction

I have read many books and articles on the "flight to the woods"—the crusade to rediscover nature and leave the complexities and pressures of urban existence behind. None that I have seen, including the Nearings' *Living the Good Life,** has been specific and detailed as to *costs*.

While a little philosophical and even emotional indulgence is perhaps inevitable and, I hope, excusable because one has strong feelings about such an adventure, my purpose is to omit nothing of the economic details encountered in our home-building project. I suspect that these considerations are uppermost in the minds of most aspirants to the good life, as it is called. They ask over and over, "What will it cost?" I did.

Accordingly, what follows is a chronicle of the delights and disasters of our experience as do-it-yourself home builders in Vermont, together with a fairly specific rundown of the economics involved in the actual construction of the house, based on careful records of our costs (see Chapter VIII).

I am still surprised at what we did. It is incredible because of the magnitude of even so modest an undertaking in terms of the work entailed. The realization of it comes only from the experience of doing it—we romantics being what we are, I suppose. It becomes believable only when the physical evidence stands there to see. Something is there on the Vermont hillside that was not there before, a presence on the landscape that certifies to the fact that a dream of 30 years was not entirely fantasy. Whether I would do it over again, knowing what I know now, is another question. But I'm very glad I did it.

The rewards are great: the sense of accomplishment, the

**Living the Good Life: How to Live Sanely and Simply in a Troubled World* by Helen Nearing and Scott Nearing, Schocken, 1970.

pride of practical achievement (look what I did!), and even more a feeling of independence nourished by doing for oneself what most would have had done for them, thereby rendering oneself less dependent on specialists. True, our bathroom door lies a little askew against its jamb, but it is functional. The latch snaps smartly into the striker and, closed, the door sits snug and square against the stops. It is, after all, the first door I ever hung.

There is in such an experience a sharpening of the sense of affinity with the past before the age of specialization assigned each of us to a particular vocational and social niche, thereby limiting our options and narrowing our perspective of a world that men have hewn with ax and hammer as well as words and technology.

The ax and hammer somehow link us to earlier ages when men were closer to nature, having fewer conveniences to distract them from nature's offerings and, by necessity, a far greater diversity of pursuits. In this sense, an affinity for nature too is awakened because I must (or wish to) struggle a little with fairly simple tools and somewhat primitive methods to put a roof over my head, just as men before me toiled with their hands and backs to harvest nature's riches from the earth and forests. For three weeks I burrowed in the earth like a mole to dig our foundation with pick and shovel, and nature yielded to my labors the result I sought—a foundation, inch by inch.

I think the endeavor brought me closer to nature and to reality although, by modern standards, my way of going about it was appallingly inefficient. A backhoe can do it in a couple of hours at far less cost in labor. The physical proximity of the open land abetted the courtship. Our stage was the whole, unblemished outdoors, the earth and air.

There is some of the same fitness in a man's building his own house that there is in a bird's building its own nest. Who knows but if men constructed their dwellings with their own hands . . . the poetic faculty would be universally developed, as birds universally sing when they are so engaged?

Henry David Thoreau, 1854

He who . . . builds a durable house . . . or so much as puts a stone seat by the wayside . . . makes a fortune which he cannot carry away with him, but which is useful to his country long afterwards.

Ralph Waldo Emerson, 1870

One of the greatest pleasures in life is to build a house for one's self . . . I notice how eager all men are in building their houses, how they linger about them . . . I see them walking pensively from one imaginary room to another, or sitting . . . wrapped in sweet reverie, upon the naked joist.

John Burroughs, 1914

Chapter I

We Buy the Land

Our five-year search; a Vermont find; financing and initial preparations for a new way of life.

The first step in the "flight to the woods" is, of course, to find the land and buy it—somehow. In our five-year search for land, we drove thousands of miles across Maine, Vermont, New York, and Pennsylvania. We subscribed to countless newspapers, combing the real estate ads and filing the likely prospects for later reference. We wrote for brokers' listings and received an avalanche of responses. Sometimes when something looked promising, I would leap into the car and drive 600 miles to inspect a property. Always there was something wrong.

For three years or so we vacationed in Maine, prying into the bays and coves along the coast from Portsmouth to Eastport. Once I saw a building lot high above the water with a panoramic view of Penobscot Bay—just a lot—price: $20,000. The Maine land boom had beaten us to it.

In October 1971 we were in Montpelier for the fall leaf festival when we saw an ad in the *Times-Argus* for 18 acres and a garage at $12,300. We took a look. It was a gorgeous piece of land with a four- or five-acre meadow sloping up to the north and east from a low point in the southwest corner. The meadow was surrounded by woodland on all three sides, the fourth being frontage on a dirt road. A small brook wound through the woods in the low corner of the land. The property rose to the crest of a hill to the east beyond the meadow and descended its far side into the woods, affording privacy to the rear. On the west there was a fine view of the Worcester mountain range, dominated by Mt. Hunger.

I was impressed. I liked the contrast of the high land and the low land flanking the meadow on the north and south. The irregular contour of the land was interesting. There wasn't a level spot on it. But as usual there were problems: in this case, no water, no septic system, and we would have to build. The garage, sitting in the northwest corner by the road, was insulated and very well built, we were assured. It was said to have cost $5,000. It had possibilities, we thought.

I filed the prospect away in the corner of my mind labeled: future business, active. As it turned out later, my wife did too. We went home to suburbia.

In the following months, I weighed the Vermont "find" from most of the usual angles. It began to grow on me, but I was waiting for a sign of my wife's reaction. I didn't want to force the issue because there was too much at stake for both of us. She said nothing.

In March 1972 we took a three-day house-hunting excursion to Ithaca, Syracuse, and Plattsburgh. Nothing we saw appealed to us.

On the way home from Plattsburgh my wife Isabel said, "You know, the more we see, the better that land in Vermont looks." From home I called the broker. Yes, the Culver Hill property was still on the market. A few days later I was in my car again, bound for Montpelier where I signed a contract to buy the place, binding the deal with $500 in "earnest money." We put our home in New City, New York, up for sale.

After some problems with the Vermont title were ironed out and intricate financial shufflings accomplished, we closed the deal towards the end of May for cash.

Basically what we did was finance the Vermont property from the proceeds of the sale of the house in New York. I found a local bank most cooperative in lending me just about whatever I asked for in the form of a short-term loan based on a so-called net worth statement. The value of our New City home was a major factor in the credit statement. I came armed with a written appraisal figure from the listing broker. At 7-1/2 percent, the bank made a couple of hundred dollars on the deal.

The place was christened "Nobody Hill" by our younger son, Walt Jr., who was strongly impressed by the contrast between the tranquil, rural setting and the clamorous suburban environment of his previous 18 years, where the bowling alley and the gang at the gas station where he worked were only five minutes away. He and his mother stayed in the house in New York until it was sold on July 7, while I commuted to Vermont to begin scratching in the ground. On each trip I carried a carful of possessions to be dumped in the garage. It is a 600-mile round trip. I lost track of the number of times I drove it.

Chapter II

We Make the Plan

Decisions, decisions—where, what and how; alternatives and discussions; architect versus do-it-yourself; what we took into account; our final solution and working plans.

You have decided to build a house. In the evolution of this resolve there are moments of truth, when decisions must be made. Where will we locate the house? In our case, will it be attached to the garage, or detached? If attached, should it be on the uphill side or on the downhill end? If detached, will we set it on the hillside back of the garage, or on the south side where the ground is most level? How big will the house be? How many rooms? What kind of a cellar? What type of roof? And so on.

First of all, you must at least know the outside dimensions of your house before you can excavate and pour the foundation on which the house will sit. The interior refinements can wait, but you should have a pretty fair idea of the basic layout before you begin. Otherwise, you may find yourself boxed in, at least economically, and unable to implement without substantial additional expense some brilliant afterthought, such as a wing extending beyond the confines of the already-poured foundation, or a recessed alcove, such as we incorporated in our plan, falling inside the foundation.

Obviously, such switches can be made, but they are expensive and time-consuming.

For weeks before the Vermont deal was closed, we fooled around with pencil and paper, sketching plan after plan and discarding them all as impractical or extravagant.

At one point, we considered an addition to the uphill side of the garage, which would have meant either joining the two roofs at their low edges, leaving a "V" angle, or raising the roof of the house so that it would meet the garage roof at its ridge, or peak, thus eliminating the "V" angle.

Both alternatives proved unacceptable, the first because in the Vermont climate you must avoid "V" angles converging at the base for fear of disastrous accumulation of ice and snow, tons of which collect on even normally downsloping roofs if the pitch is not steep enough. The second alternative presented the problem of either incorporating half of the garage roof inside the new structure where the two buildings overlapped, entailing a waste of space, or removing that half of the garage roof or raising it to form a continuous slope from the lower end of the garage roof on the far side to the peak of the addition roof, rather like a lean-to. Neither alternative was feasible for amateurs.

As we wanted to put up a building that would be compatible with the traditional New England (and especially Vermont) architecture, we thought of turning to an architect. A friend was building over on Lake Champlain, via contractor. He seduced us, briefly, with enthusiastic accounts of what an architect could come up with for a fee of only a few hundred dollars in the way of building plans that could be executed for $8,000 to $10,000. The plans he showed us were indeed imaginative, and for a time we were captivated by the notion that maybe we were overlooking some ideal design that an architect would be able to furnish immediately.

It was my impression at first that the plans our friend showed us were in the $8,000 bracket. But later it struck me that his plans did not look like $8,000 or even $10,000; neither did they look like New England. They were fairly elaborate with stuff like overhanging porches, sundecks, spiral staircase, and sunken fireplace pit.

My wife went so far as to call a local architect from our motel room in May. For $60 or $70 he would come over and inspect the property and give us some idea of what he thought might be done. The plans would come later. Somehow we never called him back.

For one thing, my elder son Steve, who has strong feelings about the merits of self-reliance, was cool to the idea of depending on an architect for our design.

"It could be interesting to work with an architect," was the way he put it. But there was clearly no enthusiasm for the prospect.

Basically I agreed with him. There was also the fact that I respect his competence in the practical manual skills of carpentry, plumbing, and wiring. I depended on Steve—the "Mountain Man," my brother calls him—entirely for the plumbing and wiring. The carpentry I could have tackled alone, but the lack of skilled assistance would have increased the sheer labor involved enormously. But I probably would not have undertaken the project at all without the Mountain Man's help.

The plan we finally came up with was the simplest of all we had considered and closely resembled one of the first rough sketches we had made, the house consisting of a simple addition to the back of the garage, having a roof raised above the garage roof level to accommodate a second-story loft and a catwalk around the open space above the living room.

The addition was to be 24 feet wide (the width of the garage) by 30 feet long. This plan eliminated the need for any

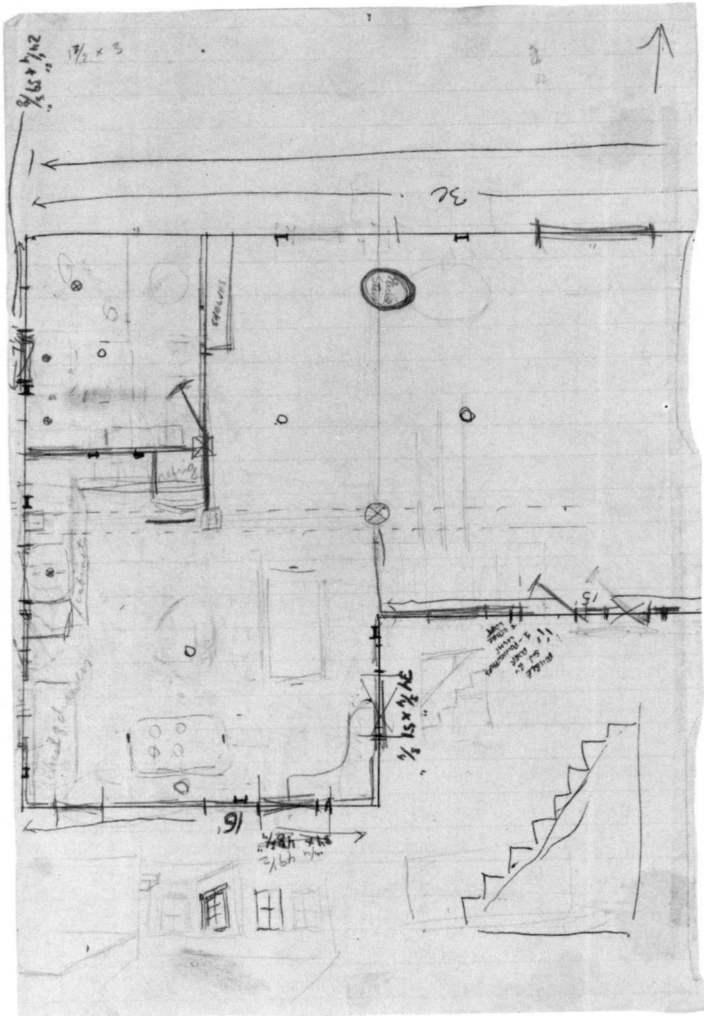

First rough layout of what was to be the final design.

fancy modification of the garage roof. We had merely to shear off the overhang in order to butt the addition right up against the rear wall of the garage. What was the back door of the garage now leads into our living room.

We pitched the house roof at an angle (35 degrees) much steeper than the garage roof for two reasons: to shed the snow more efficiently and to give the two attached components a look of authenticity. Typically in northern New England you see attached house and barn with roofs at contrasting pitches, occasionally with a woodshed intervening, but often without. And of course the virtue of this arrangement of attached buildings is that it provides ready access to house, barn, or woodshed without the need to brave the elements in inclement weather when all you need is an armful of wood.

As an added touch, we allowed for a "crazy window" lying diagonally in the vertical elevation between the garage roof below and the addition roof above on the north side, overlooking the road.

Only the south half of the house (where the bathroom and kitchen are located) is divided by a ceiling between the first and second stories, the second story of course being the loft. The north half is open from floor to second-story ceiling, an indulgence we may have to modify by sealing up the open space in some way during the coldest weather. The heat is partial to the upstairs.

Privacy being a point of emphasis, we put the front door in the back of the house, the side facing east, away from the road. Although there is a fairly respectable natural fence of trees and shrubs along the roadside, the garage is only 25 feet from the road. The door opens into a recessed, eight-by-fifteen-foot patio, as yet unlandscaped, adjoining the garage. The view looks up over the meadow to the trees beyond its crest, a panorama quite unsullied by the trappings of civilization, as my younger son discovered.

When we had settled on our basic plan, Steve meticulously converted our rough sketches into detailed working plans on graph paper, scaled 1/4 inch to the foot. He called them our "game plans." In the process of building, we deviated from these plans only where a slight adjustment would substantially simplify our task. For example, we moved the bathroom wall out perhaps six inches so that the edge of its framing would fall against one of the wall studs already in place. Otherwise, we would have had nothing to nail the framing to.

Plumbing was also taken into account in the plan and facilitated by the placement of both bathroom and kitchen at the south end on the ground floor, where the water intake pipe and the waste drain were to pass through the foundation. The septic tank and the leach field were to be located to the south of the house, the only downhill slope available for the purpose. When we poured the foundation, we left a gap in the footing to make room for the pipes to pass under the foundation below the frost line, hopefully. Locating the kitchen sink and bathroom facilities in close proximity cuts the time required to link up the plumbing and minimizes the need for long lengths of piping running the length of the house or from upstairs to the connections below. Copper piping is expensive.

The time spent on advance planning of this kind is well worthwhile in view of the time and money it can save you during the actual construction.

The Raes' final design.

Chapter III

A Family Affair

The "Mountain Man" and his crew; rewards and frustrations; trials of garage living and occasional attacks of trepidation.

I took six months off the summer of 1972 to build the house with the help of my two sons, Steve and Walt, Jr.

Kids will try any damn-fool thing, we said while our idea was incubating, but at our age . . . ? As it turned out, you don't have to be 25 to carry off such a stunt. Motive and tenacity are powerful generators of energy. I was 56 at the time of our Vermont venture, my wife was 54, the labor crew, Steve, chief contractor, 28, and Walt Jr., chief assistant, 18.

We had little experience. Steve, a candidate for a Ph. D. in physics at the University of Vermont at the time, had built a store in Stowe in the summer of 1971. He had done some concrete work, helped reroof his father-in-law's house in Stowe and had done other roofing jobs in town. An ardent believer in self-reliance, he taught himself electric wiring and plumbing and, over the years, had wired and installed plumbing in the old farmhouse his mother bought in the 1940s. He was our chief asset as far as experience goes. I congratulated myself on employing a physicist at $3 an hour.

I had done considerable manual work, most of it unrelated

to construction except for a woodland cabin I helped build 35 years ago, a simple shell without insulation or inside finish of any kind. We didn't put the windows in—I suspect we never even got around to finishing the roof. Other than that, I had built a boat, unfinished, from architect's

Stephen T. Whitney

Walter Rae (left) *with his sons Walt Jr.* (center) *and Steve. Walt Jr. is now in the Coast Guard, presently stationed in Puerto Rico. Steve is teaching physics at the University of Vermont.*

blueprints, and done a fair amount of refinishing and furniture repair as well as ship modeling.

Except for a high-school shop course, Walt Jr. had no experience. But he had brawn—and brains.

Incredibly, a house now sits on the hillside—insulated, enclosed, heated, plumbing in, and sided with weatherboard. The essentials are done, although the finish work inside goes on and probably will for some time. The cost to date: $7,820.77, plus $978 for an artesian well and $384 for a septic system, or a grand total of $9,182.77.

Economically, building one's own house is decidedly worthwhile unless all the wrong decisions are made and the do-it-yourselfer is negligent in planning and purchasing materials. The $7,820 we spent to build a house worth at least $20,000 would have spiraled, I would guess, to $17,000 or $18,000 if we had had it done by a contractor. And it is almost certain that the finished product is a better house than a contractor would have put up. Maybe not as slick, but better.

The reason is not that my sons and I know more about construction than the average contractor. The point is, being the parties with the keenest interest in the outcome and the economics involved, we put more into it in buying materials and planning construction details than could be expected of a contractor. Then, too, most contractors you would deal with buy their supplies from the dealer they know best from years of doing business with him without considering price as a major factor. You get the bill.

As a first step, Steve and I visited most of the lumberyards within a 40-mile radius to price supplies. The price of lumber varies only a few pennies a board foot, as a rule—a differential not worth bothering about unless you are buying $2,000 or $3,000 worth. Then it counts.

Quite aside from the aesthetic and practical rewards of our

undertaking, building one's own house is a way of getting acquainted with a new community. I noticed how men of the community, the well-digger and others, who set foot in our house, remarked respectfully on the finished product. Oh, the cabinetmaker had a professional word or two to say about my miter joints, which hinted that they were less than perfect, but by and large he read competence in what he saw. There was little, if any, of the skeptical reservation respecting a newcomer to the community who has not yet proved himself in something understandable and relevant to his neighbors. Our house provided a common ground.

Moreover, others, considering similar ventures of their own in the future, were eager for details as to just how we went about our building program and what tools we used. A schoolmaster who aims to put up his own house on a sabbatical questioned me intently along these lines. Had we used a jointer? A table saw? I told him the only power tools we used were portable skill saws, drills, and jigsaw, which no doubt accounts for the plain, rough look of our product. For the rest, I told him, we relied on the simple hand tools—pick, shovel, saw, hammer, chisel, screwdriver, level, plumb line, and so one. He followed my words with the earnest attention and gravity a believer gives to Gospel.

On the minus side, the drawbacks are formidable. Frustrations, aggravations, disappointments, and fiscal surprises array themselves in an endless procession to taunt the embattled entrepreneur with the implacable aim, it sometimes seems, of dissuading him from his ridiculous purpose.

There were the frustrations of late deliveries, partially filled orders, out-of-stock problems on needed materials, delays due to illness and accident, inclement weather; the aggravations of leaky toilet plumbing and faulty water-heater connections, or a sheetrock nail impaling an electric wire in the ceiling; the hazard of mud on the sodden, denuded building site, and the constant soaking of the partially built structure by the Vermont monsoons of 1972; the dilemma of where to dump the garbage and refuse; and the necessity of toting all our water from a neighbor's up the hill—a series of circumstances imposing all but impossible demands on time and pocketbook. A 24-hour day often seemed but a moment in which one barely had time to grasp a hammer, much less do anything with it. And I came to wonder whether any mechanical contrivance is really made to function properly. You install a water pump. Then you have to put in a water softener to make the pump work right. Another $250 of unanticipated expense. The entrepreneur survives only by a determination equally implacable.

We could not enjoy the luxury of a leisurely pace, as Helen and Scott Nearing could. There was always an eye on the calendar, the pressure of time marching on. Our building schedule originally was something like 18 weeks, June to mid-September. The Nearings prudently allotted a decade or more to the development of their more ambitious compound in southern Vermont in the 1930s.

The emotional drain exacted by the anxiety of whether we were going to make it before winter set in almost floored me at times. And our shrinking exchequer took its own toll.

For three months we lived in the garage, the only building that came with our 18-acre package. We camped there—Isabel, Walt, and I—amid stacks of building materials, the whine of power drills, the angry snarl of electric saws, and dust—sawdust and, worse, sheetrock dust, a sort of gray pervasive haze undaunted by broom or mop. Our worldly possessions were piled in a corner, shrouded in dust.

We cooked on hot plates, fetching provisions from an outdoor refrigerator I fashioned by digging a pit and lining it with burlap, saturated nightly to help retain the cool night air. A garbage can lid held in place over the pit by a rock discouraged animal poachers. Isabel was the camp cook. (Concession No. 1: We later abandoned the outdoor arrangement and bought a refrigerator for $25. I couldn't have gotten to the pit in winter, anyway.)

The circumstances of this living arrangement were such as sometimes to impose a dreary emotional strain on family relationships. Walt bedded down in the garage with us after two nights of isolation on Nobody Hill in a tent we set up for him. Cost: $82 at Sears. Pretty expensive accommodations at $41 a night.

"There's no noise on Nobody Hill," he complained. "The only thing there is stars."

Isabel is an orderly person. I am not. Up to a point, I am more or less indifferent to order. Our quarters were close. As the months wore on, there were demands to put things away, hang up clothes, remove shoes and socks from avenues of traffic, fetch water, and take out the garbage, cans and bottles. I'm afraid I wasn't always responsive or even sensitive to her appeals. In retrospect, they were perfectly reasonable, but they began to pall, and it annoyed her that her pleas for cooperation often were ignored. This put her in the humiliating position of having to repeat herself to get my attention.

I sometimes felt her demands were excessive and even irrational but, even when I did not, I chafed at the need to spend time on housekeeping chores even though I recognized the need for some kind of order.

Either I was so focused on the building project or, in Burroughs' words, so busy sitting on a naked joist "wrapped in sweet reverie" that my wife's adjurations failed to get through to me. I was busy with the physical work of building too, and I sometimes thought she could have modified her need for order in view of the circumstances. We both, I suppose, could have given a little.

Among the feats I performed with pick and shovel was the creation of a woodland privy up beyond the stone wall in a fern patch. But my wife wearied of the daily pilgrimages to the woods, especially when the temperature dipped to 32 degrees in October before we had the toilet in and operating. This crisis, like others, passed, and tensions relaxed perceptibly after the toilet was in and working, albeit with a stubborn leak that later yielded to Steve's ministrations with sealer.

Later, the completion of our kitchen cabinets figured as another milestone in Isabel's adjustment. She was finally able to get our larder and kitchen utensils off the floor where they had reposed for seven months, first in the garage and then in the house. I was relieved to get the stuff off the floor, too. It had begun to get even to me.

We bathed in a mountain stream two miles or so down the hill, or in a porcelain tub in the garage, with the shades drawn, when foot traffic permitted. Always there was the portage of water, to be imported down the hill from our neighbor's, or to be discarded in the field. When a hot bath became a matter of real urgency, we took refuge at our favorite motel.

After an investment of $300 in three handsome wood stoves restored to pristine perfection by an expert craftsman, the disappointing discovery that a member of the family was allergic to wood smoke, at least in the volume emitted by a cook stove, led to the purchase of an electric range. Cost: $209, plus $30.06 for installation, bringing our total investment in stoves to $550. The wood stoves are idle most of the time.

On top of this came the realization that one of the basic expectations of our flight to the woods—that of heating chiefly with wood—must be abandoned. In its place we substituted electric baseboard heat which I had planned to

Stephen T. Whitney

Isabel Rae with her alto recorder. The portrait on the wall is of her grandmother.

install anyway, but only as a back-up system. Baseboard heat is clean and efficient, but the cost is steep.

Such calamities are probably inevitable in a self-moved venture of this kind into the relatively unknown or at least unfamiliar. Everything cannot be foreseen, nor can every pitfall be anticipated. But one way or another we persevered in our project. The will has powerful resources, although there were times when I brooded, even aloud, "Whose idea was it to build this damned house anyway?" My sons laughed.

"We were unrealistic," we exclaimed at one point. "The dream, for both of us, was part fantasy. At our age, it's just too much of a gamble."

Maybe so. But the mood passed. I never really lost confidence in our ability to complete the undertaking. And the fact is that the damn-fool thing is *there*—the work of our own hands. We are living in it, agreeably comfortable, and the view of Mt. Hunger in the Worcester Range to the west is magnificent and ever-changing.

The west wall of the Raes' house. ▶

A FAMILY AFFAIR

Stephen T. Whitney

Chapter IV

We Start From Scratch

Digging ditches; using the transit to level our trench; costs; footing and foundation wall; making the frames and pouring concrete; a brush with disaster, and something on the subject of anchor bolts.

Armed with our carefully formulated game plan, we began digging for the foundation in June. We dug and dug and dug.

First, of course, we staked out the lines of our house to the proper dimensions with Steve's $2 homemade transit, but the dominant activity at that stage in our enterprise was digging.

"You go down before you go up," Steve commented.

Where there had been nothing but grass, a trench began to appear. The mounds of earth we moved piled up along the trench.

It rained persistently almost every day in June, July and half of August. We dug in the rain, the trench awash and slippery with mud. Though we gave in only occasionally to the inhospitable weather, it was often enough to slow us up.

In June I was living alone in the garage while my wife tied up the New York end of the operation. I cooked, after a fashion, on the hot plates in the confusion of an unmade bed, dirty clothes, grime, and clutter. The mud we tracked into the place stayed there until it dried and turned to dust. It was almost impossible to sweep the concrete floor, except for random patches, because of the litter lying around. Once a week maybe I washed the dishes and made the bed.

"I'm going to get organized one of these days," I assured my disbelieving son. But I never did, until my wife arrived. Among other things, I was so exclusively focused on the

You go down before you go up. Walter Rae "scratching in the ground" with a pickaxe.

house and the urgency of making headway that I begrudged the time it took to wash my hands—when I did.

At night, after Steve had gone home to Stowe, I was usually too tired to cook a meal. Often there wasn't a clean dish available anyway, so I would drive to town to eat. Bedtime was 8:30 or 9:00; sometimes as late as 10:00.

We planned a trench two feet wide and some three feet deep, on the average, to accommodate a footing eight by ten inches and a foundation wall eight inches wide and two feet tall on top of it. The trench should be considerably wider than the footing to allow both working room and space for the frames that must be built to hold the concrete in place

The foundation trench nearly ready for the footing forms.

when it is poured. The frames are of course removed after the concrete has set.

As we neared the desired depth, we checked periodically with the ingenious transit which Steve had fashioned from a discarded tripod, a piece of plywood, a pair of one-dollar levels, a short piece of pipe, and a sheet of aluminum he calibrated with a sharp-pointed tool to indicate the necessary angle readings. He claims it is accurate within an eighth to a tenth of a degree, or about one-half inch over 30 feet, and perfectly adequate for home construction purposes. And I believe him.

Because you want your house to be level, the foundation must be level regardless of the lay of the ground on which it sits. The considerable slope of our site meant that our trench would have to be deeper in the high ground than in the low ground, so that the footing would be level.

When you have reached the correct depth at one spot, you check with the transit all around to find out whether the depth from a horizontal line is uniform. This is done by placing a straight stick in the trench and marking it with a piece of chalk to indicate the correct depth. The mark is visible to the man at the transit through his telescope—in this case Steve's piece of pipe which he set exactly horizontal with the aid of his dollar levels. With the transit stationary so that the elevation doesn't change, the man with the stick places it upright in the trench at various points. If the chalk marks all show up in the sight of the transit, your trench is level. If your chalk mark is too high, your trench is too high in that spot, and you must go back to digging. If the chalk mark is low, your trench is low, and you must fill.

The day came when we had gone far enough in three weeks of digging. The trench was the right depth and checked out level.

We did not dig a cellar—too much work, added expense,

and not really necessary for insulation purposes. After our foundation was in, we excavated sufficiently to provide a minimal crawl space under the entire floor. But we went down about two feet under the south end of the addition to make it easier to get at the plumbing.

At this point we had spent $478.91—$5 for a building permit, $249 wages to Steve, $220.57 for lumber and plywood for the footing and foundation wall frames, and $4.34 for nails and wire to tie up the reinforcement bars in the foundation framing. The figure doesn't include my labor. But discounting this factor and conservatively speaking, I would say it cost us three times what a man with a backhoe would have charged for the excavation work alone. The materials would have cost the same.

We did not get an estimate for this, the initial phase of our project, because, I imagine, Steve and I sort of by tacit agreement shared a determination to do it ourselves from the beginning.

However, judging by the excavation work we had done later for the septic system for $135, I'm sure that mechanical excavation for our foundation wouldn't have cost any more than $150, indeed closer to $100.

In other words, for an additional $225 or so, plus close to three weeks' investment in time, I have the perverse satisfaction of living on a foundation I dug myself. Perhaps it's worth it. But it adds up to economic folly beyond a doubt, the only such extravagance I think we were guilty of. If I were doing it again, I would call in the backhoe.

When we had checked out the trench, we went ahead and built the forms for the footing from green lumber along the bottom of the trench. We braced it with stakes driven into the ground on each side and with stringers nailed across the top edges of the framing boards as a precaution against the forms buckling under the weight of liquid concrete.

Among our discoveries from our pricing expeditions was a supplier of green, rough-cut lumber at around eleven and a half cents a board foot, just about half the going price of kiln-dried lumber at the time. Lumber and plywood prices went up during our project, fortunately after we had done most of our bulk purchasing.

It is obviously a mistake to use green lumber for any permanent installation—floors, studs, siding, and so on—because of future shrinkage as the timber dries. You flirt with the prospect of unwanted, built-in structural distortion and unplanned air-conditioning from cracks opening up in walls and floors, with disastrous impact on your heat-loss factor. But there is nothing wrong with green lumber for such temporary purposes as concrete retaining forms. Our first lumber purchase was green one-by-ten-inch boards for the forms we made for the footings—at a considerable savings.

Our concrete, five cubic yards of it, was delivered by mixer on the appointed day and poured into the forms. Cost: $144.25, including the reinforcement bars.

We removed the framing the next day, and applied ourselves to making the more elaborate forms for the foundation wall. We made them in eight-foot sections (the length of a sheet of plywood) of plywood and two-by-four framing, installing them end to end on top of the footing. We braced them with vertical and diagonal stakes, stringers across the tops, huge rocks at the base, and reinforcements added at the corners—in every way we could think of—against the weight of the concrete on the inside of the taller, narrower forms. This time, we figured, the center of gravity would be higher than it was in the lower, wider footing frames, and the pressure greater.

The mixer arrived, this time with seven cubic yards of the stuff, and the pouring began. It started to rain hard on the bare, trodden earth, converting the hard-packed, clay-laden ground where we worked into a perilous slick of slippery mud in minutes. Then, as we guided the liquid concrete down the chute and spread it out in the forms, we heard the sharp snap of a timber giving under tremendous

The completed concrete footing (forms removed), ready to receive the foundation wall forms.

pressure. Another followed. In each rending tear of timber, I physically sensed the awesome weight pushing at our frames.

We leapt to avert the impending disaster. In spite of our pains, the south wall frames had started to give—first one, then another. The driver jumped from his rig. Steve seized a crowbar, and I clutched the biggest rock I could find to heave against the base of the spreading frame.

"Not there. Here," Steve ordered. There was a sharp edge in his voice, his honor being at stake. I obeyed, trembling before the vision of a $700 investment, and by now over three weeks of labor, dissolving on the ground if the frames let go.

We battled desperately to shore up our creaking frames with rocks, timbers, and crowbars. I thrust a two-by-four into the trench against the base of a frame and pulled it against a rock one of us had thrown in to get leverage. I pulled on it for all I was worth. Steve was shoveling some of the escaped concrete back against one of the snapped frames. It was still raining—pouring. Twice I slipped and fell in the slime, once skidding into the trench and giving my leg a nasty wrench.

I don't know how long the nightmare went on, but we didn't relax our counterattack while the emergency lasted. I remember the driver, 30 years my junior, taking a shovel from me and carrying on my attempt to bank up the earth against a weak spot.

The rending subsided, but the ordeal wasn't over. We still didn't have our wall, but we seemed to have succeeded in shoring up the two buckled frames sufficiently so that, though misshapen, they held where they were.

The driver wiped the perspiration from his face.

"Whew," he said, "if they'd gone down, the only way you can get rid of that stuff is with a jackhammer."

With the utmost vigilance, we surveyed the other forms for

vulnerable points and reinforced them with additional timbers, rocks and stringers. The concrete pouring resumed. I stood there with a crowbar, listening to the sounds of stress, all my senses focused on the forms, straining to detect the slightest sign of weakness. The concrete oozed into the upright compartments. The forms held!

The driving rain peppered the ground and splashed on the wet concrete. The trenches were filling with water, so Steve dug a drainage ditch at the low point at the south end of the foundation to carry the water off.

Pouring concrete is one of those jobs you must stick with until the job is done. In my innocence, I had always thought of the phenomenon as simply a matter of installing the forms properly, pouring the concrete into them, and waiting for it to set, perhaps with a cold beer in your hand.

Instead, I discovered we had about two hours of work ahead of us, exhausted as we were. The anchor bolts, about

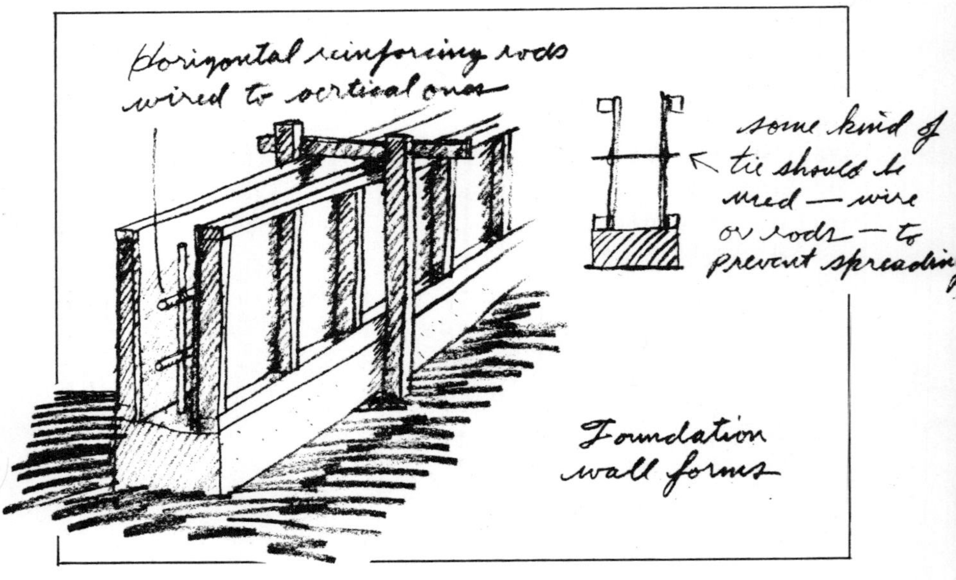

36 of them, had to be embedded, thread upward, in the wet concrete. We had to smooth the concrete with mason's trowels all around, making sure the concrete filled the forms flush with the top, and that it was level and free of air holes. The driver left us a wheelbarrow full of concrete to fill in the low spots. We checked with a level as we went along.

Finally, we covered the entire wall with 150 feet of plastic sheeting to protect it from excessive wetting by the rain. This precaution is standard practice anyway to retard the speed of setting. If concrete sets too rapidly, it may crack and develop a flaky surface that chips off.

There is another way, however, for those who would prefer to avoid such a harrowing experience.

Purveyors of ready-mixed concrete supply steel forms in standard sizes which they will set up for the home builder for an additional charge. No doubt they are stronger than wooden frames and fairly foolproof—provided they are set

Steve checking the forms before arrival of the concrete mixer.

up right. Our driver, for example, told us of a disastrous experience one of his co-workers had had. In setting up steel forms for a customer, he neglected to attach two adjoining forms securely. When the concrete was poured, the seam parted, and the concrete oozed out on the ground, botching the job irretrievably.

We had considered concrete-block construction of our foundation, basically because of its manageability. Small quantities of concrete at a time suffice, as the blocks are built up one at a time on top of each other. Thus, the necessity of pouring a whole wall all at once is avoided. The time element was the deciding factor here. The leisurely concrete-block work would have taken much longer. Who knows, we might still be at it—so we marshalled our courage and decided to pour the walls and be done with it.

In any case, the rule is: respect the weight of liquid concrete and make your frames strong.

That night, after working until 8 P.M., I sat on the edge of the bed in the clutter of the garage, stupefied by exhaustion, bemused by our brush with catastrophe, and marveling at the wonder and perils of pouring concrete. I felt rather like a man returned unexpectedly from the gallows by virtue of an eleventh-hour reprieve who has not yet quite grasped the fact that he is alive. I remember making a conscious decision to take off my wet shoes and socks, and being vaguely surprised to find myself sitting there a half hour later not having made a move to carry out my resolve.

Cost to date: $768.80, labor and materials. We had our foundation. It had middle-age spread in a couple of places, but it was a straight, strong, and level base on which to set our house.

Steve did get an estimate for a complete foundation job, including a half cellar, for something like $1,150, exclusive of the cost of excavation. We would have gotten a little more, but at a third again the cost. And there are always contingent problems. If you have a cellar, then you need a cellar door and maybe a window or two at additional cost.

"I sat on the edge of the bed . . . stupefied . . ."

Chapter V

We Build the Frame

Onwards and upwards—the sill, floor joists, and subfloor; the wall frames; "16 on center" rule; studs; top plate; a successful foray for barn beams; rafters and ridgepole; gable studs and a crazy window.

The first step, after your foundation is done, is bolting the sill to the anchor bolts protruding from the top of the foundation wall.

The sill, despite its lowly position in the elevation of your house, carries the distinction of being its first permanent wooden feature. It is a two-by-eight-inch plank lying broadside on top of the foundation wall all around. The nuts of course must be countersunk into the sill so that the bottom edges of the wall frames will sit flush on top of the sill. Sill timbers, like most other horizontal features, are put down with the aid of a chalk line to insure that they are straight and square at the corners. Also, like many other features, horizontal and vertical, they are constantly checked with a level before being bolted permanently into place. If necessary, slight adjustments can and should be made by shimming.

Our next order of business was to put in the floor joists, two-by-eights lying on edge across the foundation and resting on the sill.

In building our foundation, we ran one wall up the center

from the south end. This was done to satisfy the construction principle of not running floor joists more than 16 feet without support from below. If the span is greater, the floor tends to sag or bounce. Since our building was 24 feet wide, we needed additional support up to the point where the recessed patio extended out from the garage on the east side. There the span was exactly 16 feet inasmuch as the patio was eight feet deep.

On top of the joists we laid the plywood subfloor, or most of it. The south end was left open for the time being so Steve would have access to the area where the plumbing was to go in. The subfloor gave us a convenient flat surface on which to work, permitting the removal of a good part of our operation from the garage where we lived, or tried to. My wife was delighted to see the stuff go.

In building our wall frames, we followed the standard construction principle of spacing the studs "16 on center"—that is, 16 inches from the center of one stud to the center of the next stud. The same principle was applied to putting down the floor joists.

In general and where possible, everything is laid out in multiples of four feet: 4, 8, 16, 32, and so on. The reason is that most building materials (plywood, sheetrock, paneling, etc.) are made in four-foot widths and are usually eight feet long. If the framing is built to accommodate these measurements and the "16 on center" principle is followed, it saves an enormous quantity of work in cutting and fitting materials to match odd-sized frames.

For example, if you are laying four-by-eight plywood sheets across the floor joists, as we did, the eight-foot sheets hopefully will fall exactly along the center line of one of the joists. The next sheet will butt up against the first on the joist, and you have something to nail both sheets to where they meet.

Similarly, if you are laying plywood or sheetrock parallel to the studs as you do on the wall frames, the same convenience is provided by the "16 on center" spacing. The adjoining edges of plywood or sheetrock will butt along the center of the studs where both can be nailed.

Another advantage of the "16 on center" rule is that you

Bolting the sills (sill plates) to the foundation walls.

always know where the joists or studs you have covered are when it comes to nailing. You can mark your nailing line simply by taking your tape and measuring off 16, 32 and 48 inches from the edge of your plywood or sheetrock.

An even surer way to get the nailing lines accurate is to put little chalk marks on the frames and on the floor, above and below, to indicate the position of the studs. Then, after the sheetrock is tacked on, all you have to do is run a line between the two marks, and you can't go wrong.

Of course these convenient rules of thumb often must be modified to accommodate windows and doors and walls of different sizes and lengths. Our walls, running north and

Excavation of crawl space complete. Small openings are for ventilation. Note "middle-aged spread" in wall (left).

Installing the subfloor over the joists. Where the span is 24 feet, the joists meet at the center wall.

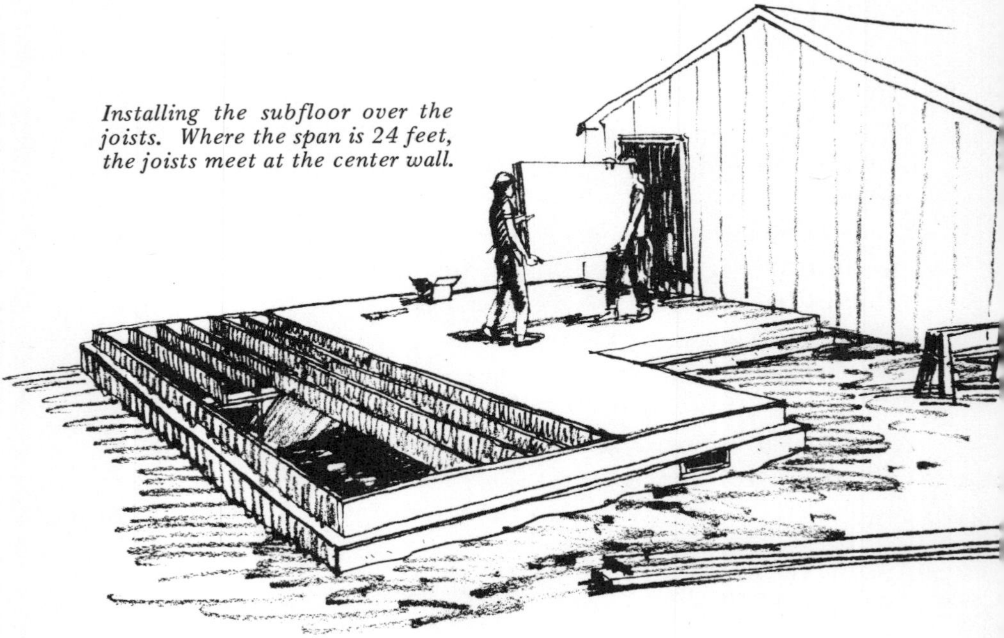

south, are in fact 31 (not 30) feet long. In either case, our frames would not have come out to the correct measurement if they had all been four feet wide. So we simply made the last one in the line narrower. Neither is there any law which prohibits spacing studs closer or farther apart if you have to. In adjusting spacing of joists or studs, it is usually better to move them closer together rather than farther apart when strength is a consideration (as in floors and walls).

Before you build your wall frames, you must know where and how big your windows and doors are going to be because their spaces must be built into your frames. If you reach this stage without having planned your window and door layout thoroughly, you will have to do it now before proceeding with the framing. We had bought all of our windows and doors, except for some of the storm windows, by the time we got to this point.

We made our wall frames eight feet wide by 11′ 1″ tall,

"16 on center," which gave us a 7′ 9″ ceiling downstairs and 27″ clearance upstairs to the point where the walls meet the roof rafters. The thickness of the floors upstairs and down accounts for the "lost" 13 inches.

The 35-degree slope of our roof provides plenty of headroom for a 14′ x 15′ loft. For storage, I built closets running the full length of the loft floor area, north to south, to make use of the dead space where the roof is too low for headroom. On the west wall I cut into the floor area somewhat more by building a deeper walk-in closet.

Our wall frames were built flat on our plywood floor and raised into position as we finished them. They were nailed with three-inch common nails to the sill and to each other. Where the frames joined vertically, we had double studs to which to nail the outside edges of the plywood and sheetrock.

Being tall and quite heavy, the frames—the skeleton of

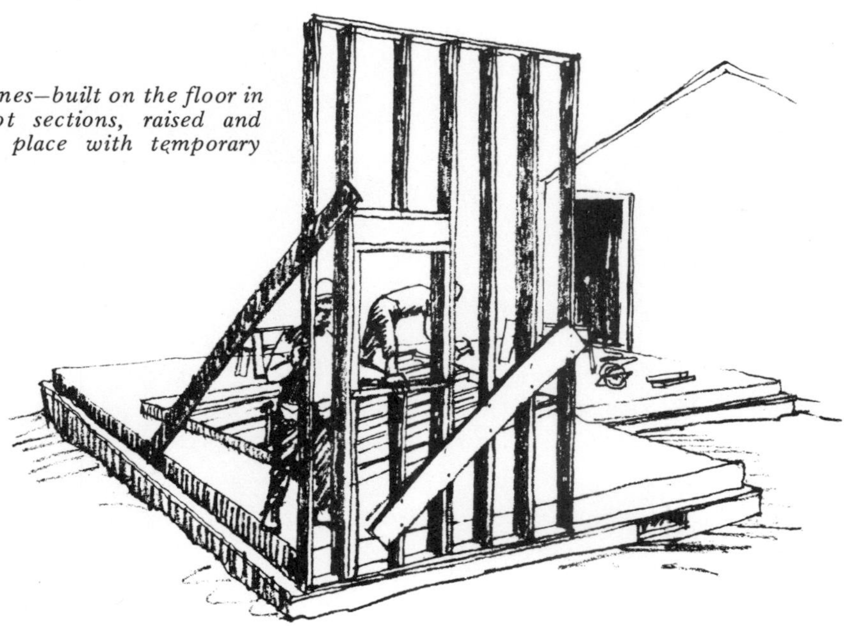

Wall frames—built on the floor in eight-foot sections, raised and held in place with temporary braces.

the house—were quite shaky before any horizontal support was installed. Strips of scrap wood tacked diagonally across the corners temporarily held the frames more or less steady and kept them square.

When the frames were all in place, we had an impressive skeleton rising over 11 feet from the foundation. We were getting up there, out of the sodden trenches where we started. The frames were all tied together by a top plate nailed along the tops of the frames.

Next, two-by-fours were nailed on the inside of the frames along the length of the house on both sides to support the upstairs floor joists.

We wanted some hand-hewn barn beams with which to adorn the interior of our mansion and to frame the exterior of the recessed patio. Fifteen to twenty years ago you could have had them for a trifle, but barn beams have been "discovered" in Vermont. The going price is one dollar a running

Three wall-frame sections in place.

foot. I approached one gentleman who was stripping a collapsed barn on his property.

"I'll take a dollar a foot for the beams that are left," he said. "Then everybody's happy."

I said I'd think it over.

The next man I heard about with barn beams was another story. He didn't say anything about his price, but just remarked that it was a shame nobody had used the beams from the barn he had torn down, because they were rotting in the undergrowth. He wouldn't have them in his house—"worms, probably," but conceded that some folks liked them.

"What do you want for them?" I asked him.

"Dunno what they're worth. What'll you offer?"

"Dunno," I said. "Let's have a look at them. What kind of condition are they in?"

"Dunno," he said. "Laying here in the grass this way, you can't hardly see 'em. Now here's a pretty one."

Wall framing completed.

Beating the undergrowth from around the beams with a crowbar, he battered the beam with the tool.

"Solid as a rock. No rot here. How many you want?"

"Dunno," I said. "Depends on the price. Sixty, maybe eighty feet."

"Hell," he said. "Why don't you take 'em all? They ain't no use to anybody, laying in the grass this way."

"I don't know how I'd get them over to my place," I said. "I guess I would take a few at a time in the wagon."

We had bought a large, nine-passenger station wagon for carting lumber and materials.

"Hell," he said. "I'd cart them over in my new pick-up. Nothing to it. We could load 'em up right now. I ain't got nothing else to do excepting talk to the women, being Sunday."

We were standing in a field behind his house in the hot July sun, shifting our weight from one foot to the other as the conversation wore on. I ran out of cigarettes.

After a time, having reached an impasse, I passed the burden of the banter over to Steve who is more practiced at this type of exchange than I am—and enjoys it, I think. It is really a game of wits. You must listen carefully to what is said for a clue. And, finding a clue, press to capitalize on it after a decent time lapse.

For a moment I thought of offering the gentleman $25 and taking the 60 or 80 feet he seemed eager to unload, or whatever he would part with for the price. But I was not certain yet whether he was building up a sales pitch at the going rate. Also, suspecting he was quite aware of the going price, I hesitated to insult him by offering him a ridiculous sum which both of us knew was out of line.

Finally Steve asked him how much he'd charge to deliver the beams. Nothing had yet been resolved as to price.

"Dunno. Ten dollars maybe," he said. "Hell, delivering

'em is worth more than the beams are."

That was the clue but, according to the rules of the game, the coup de grace must wait. Some minutes later after an extended exchange of noncommittal pleasantries, Steve said in a sort of drawl, "Well, you said it's worth more to deliver them than the beams are worth. If you will deliver them for ten dollars, how about eight dollars for the beams?"

"Fair enough," our man said happily. "Glad to have someone have them that wants them. They ain't no use laying here. Don't do 'em no good either. I hate to see 'em half growed over. Should have put 'em up off the ground. Meant to, too. Now if I can get this blasted tractor started . . ."

The bargain struck, I realized that he would have sold me all the beams he had—200 to 300 feet—for $8. We loaded the beams into his pickup using the shovel of the tractor to hoist them into the body. The upshot of the matter was that I got 70 to 80 feet of beams, carted to my place, for $18, about a quarter of the going price, which normally would not include delivery.

Several weeks later we decided we could use more, so I repeated the deal, this time far more expeditiously. But we had to leave one of the beams I selected behind. Though it was a beauty, the beam was too long and heavy to balance right in the truck. He attempted to load it anyway.

"Hell," he said, "we'll put her on. We can set a couple of them shorter ones on the end of it next to the cab to hold her down. She won't go nowhere."

I warned him it was too long and heavy for the tailgate where a good deal of the weight would fall.

"Nope," he said. "We'll get her. Just you wait."

He picked the beam up in the scoop of his tractor and tried to balance it on the load so the end of it wouldn't drag on the ground. It didn't work. We had to leave it. He said, damn it, he'd bring it over some time, somehow.

". . . riding on the business end of an inverted jackhammer."

A week or so later we were sitting on our hillside on a lovely August evening drinking beer when our man's truck rolled into the yard below followed by an impressive cloud of billowing dust. His wife was at the wheel. Our man was perched on the end of the lone beam in the body of the truck being rudely jolted up and down as though he were riding on the business end of an inverted jackhammer. He had the beam lashed down with chains but, in spite of his pains, the tail end of the huge beam was dragging on the ground. It had been all the way, it was obvious.

Our man jumped down and shook himself together from the drubbing he had taken.

"Well," he said. "There 'tis. Smoothest ride you ever saw. I told her that beam weren't going nowhere, especially with me on top of her. Figured an extra hundred and thirty would help weight her down just nice."

He laughed. He weighed more like 180, which was just about as effective as 130 pounds in holding that beam down.

When we pulled it off the truck, I saw that the tailgate was bent.

"Hell," he said, noticing that I had seen it, "it ain't nothing. It'll snap right back."

I offered to pay him.

"Nope," he said. "That ain't the bargain. I said I'd bring it and there she is. Besides, I saw you was disappointed, leaving that beam behind; saw you kind of had your eye on it. Glad that somebody has 'em that appreciates 'em."

He did accept a cold can of beer.

"Yes, sir," he said. "That's my weakness, beer is."

I not only got a bargain but also a pleasantly warm encounter with a very decent guy.

Our barn beams went in, five vertical posts and two horizontal beams resting on top of them. One of the beams runs across the room under the edge of the loft and in under the top step of the stairs. The other, at a right angle to it, runs under the center of the loft floor, supporting the joists.

The beams, besides being functional as supports for the loft, also serve to delineate the kitchen and living areas. For me, the beauty of the massive, old gray timbers is much enhanced by their function—here the same as in the barn they came from. I am not partial to elaborate architectural frills that have no purpose other than ornamentation. Indeed the charm of early New England architecture owes much to the simplicity of its functional design and the ingenuity of the artisans who did the work with axes, adzes, and wooden pegs.

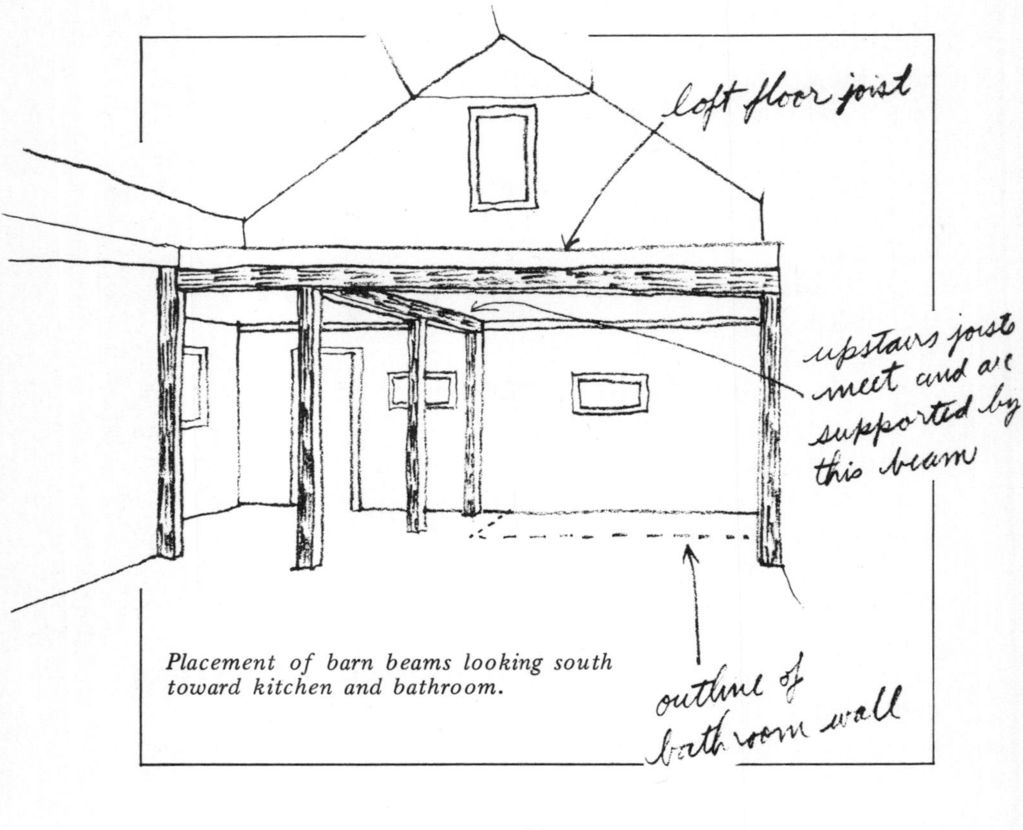

Placement of barn beams looking south toward kitchen and bathroom.

Barn beams were used on the exterior to outline the recessed entrance patio. (The fish is a sand sculpture by a neighbor.)

In any case, the beams in, we proceeded with the upstairs floor joists, two-by-eights laid in the approved "16 on center" manner from our footings on the walls. The joists met on the beam in the center under the loft floor.

In the north half of the upstairs, we modified the framing to leave the space over the living area open. A four-foot catwalk extends from the loft along the east and north sides of the house over the patio ceiling and under the crazy window in the north wall adjoining the garage roof. The loft and the catwalk were floored over with half-inch CDX grade plywood. Steve felt I should have bought five-eighths-inch plywood for the purpose to give the floor greater rigidity, but I haven't noticed any give or bounce in the floor, and I weigh almost 190.

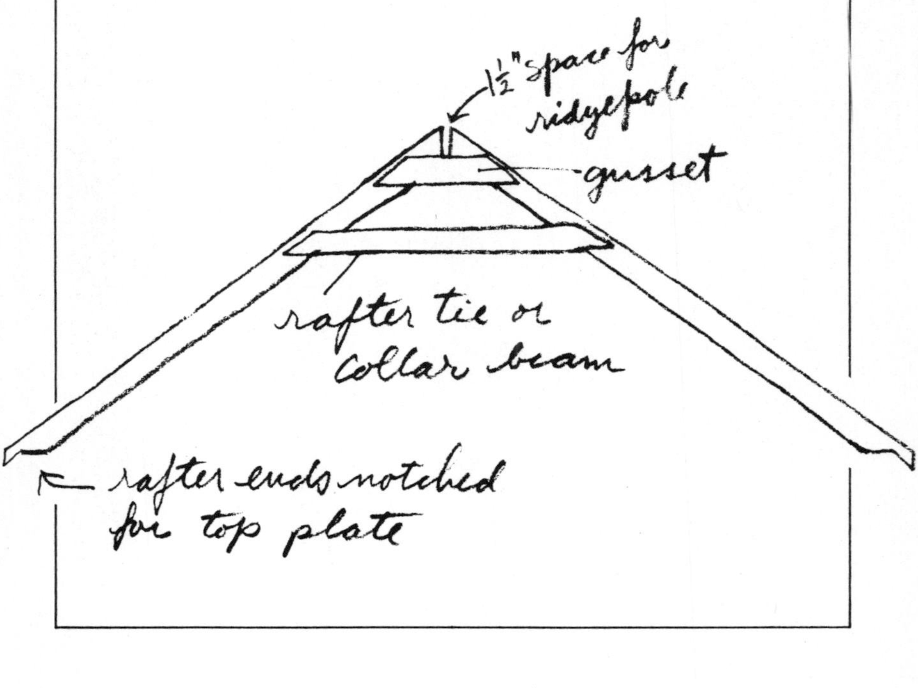

With the help, as he put it, of a little Euclidean geometry, Steve drew a full-scale pattern for the roof rafters on the loft floor. They were made in pairs, notched at the lower ends to fit flush over the top plate and overhang it by four or five inches. The top ends, cut to the proper angle, were attached with plywood gussets so as to leave a gap of one and one-half inches between the tips of the rafters to receive the ridgepole when all of the rafters were in place. Each pair was also attached by a horizontal spreader (collar beam) about half way up. The spreaders formed the framing for the loft ceiling and were later covered with sheetrock.

Steve's geometry was so accurate that when we hoisted the first pair of rafters into place, they exactly spanned the 24 feet from side to side and sat flush down on top of the plate.

Steve nailed the temporary brace into the garage roof.

The angle cuts on the lower ends came out exactly vertical at the overhang, and the angle cuts at the peak were vertical and of course parallel, forming a perfectly rectangular notch in which to drop the ridgepole. Even Steve, phlegmatic as he is, was pleased with the results.

"Not bad for highway work," he conceded.

Even with Euclid's help, carpentry involving angle cuts, except possibly for simple rectangular framing where all the cuts are 45 degrees, is a little hairy—and every cut made in those rafter timbers was an angle cut. First you must calculate your angles and then cut them correctly, and in the right direction. It is easy to goof. Nothing easier!

Ridgepole is in.

But since the first pair of rafters fit perfectly, all the rest did too, the pattern for all of them being the same.

If our wall frames at first were shaky, the rafter assemblies really wavered precariously on top of our frame, there being no support at all except nails through the lower ends into the plate, 24 feet apart; nothing on top. They squeaked and swayed menacingly in the wind. We supported them with temporary diagonal braces nailed across the rafters and even into the garage roof (from the spreader of the first pair of rafters), the only place this last brace could be anchored.

"Hey," I said to Steve, "how come you're driving nails through the roof of my garage?"

"You don't want your house to blow down, do you? You can fix it with a little roofing cement," he said.

I conceded his point.

To complete our framing, we installed the gable studs between the top wall plate and the rafters at both gable ends, allowing for our window openings. We placed these one by one rather than making frames in sections as we had for the wall frames. Doing it this way was really easier since these studs are all of different lengths, increasing as you approach the ridgepole. And probably a more accurate fit was achieved by measuring each stud separately. On the south end, the tops of the studs were cut to the angle of the roof slope to fit snug under the end rafters.

The north end where the house attached to the garage was trickier. Here we nailed a two-by-six-inch plate along the top edge of the garage roof and ran our studs from the plate to the rafter above. But in this case we had two angles to contend with, the angle of the garage roof below and the angle of the house roof above.

Actually, we had a couple of more angles to cope with on the north end, since I had decided to install one of those so-called "crazy windows" that lie diagonally under the roof

Stephen T. Whitney

Exterior and interior view of the crazy window in place. It now illuminates Walter Rae's writing desk on the catwalk.

Stephen T. Whitney

overhang. They are fairly common in Vermont houses and, I think, picturesque.

When I came to grips with it, the framing for this was easier than I expected. My crazy window lay parallel to the roof of the house, which determined the angle of the two sides. To get the angle at the ends, I merely had to take a square and get a right angle with the parallel lines at both ends. The framing entailed a lot of cutting of short studs that ran up under and above the window frame. Each had to be cut to fit the angles at either end. I masterminded the project, with Walt's help and advice; we did a pretty good job, I thought. I was quite proud of it.

Curiously enough, in discussing the origin of the crazy window with several Vermonters, no one I talked to seemed to know. The best guess seemed to be that somewhere along the line a carpenter must have had a window he wanted to use under a roof where two buildings joined, and found it wouldn't fit vertically. So he simply tilted it until it did fit and did his framing accordingly. Academic research might prove differently, but this seems reasonable to me and quite in keeping with Yankee ingenuity.

Chapter VI

We Sheathe The Frame

The roof goes on; Steve's ingenious scaffold; plywood sheathing; installing the windows; hanging doors, and some trouble with a latch-lock assembly.

It was August. Our skeleton was complete. Although we were somewhat behind schedule, our progress was encouraging nonetheless. We were "right up there as far as we were going," in Steve's words.

By now our wall frames were fairly rigid, held together by the upstairs floor joists stretching between them. But even more strength is added to the frame when the outer sheathing of plywood is applied. It knits it all together. In any case, the windows and doors cannot be installed properly until the plywood sheathing is on. For one thing, the window casings, when they come with the windows, are nailed into the frame through the plywood. Otherwise the windows would be recessed too deep and exposed to seepage of rain water. That's bad.

We scattered our forces. Steve applied himself to the roof, Walt tackled the outer sheathing, and I busied myself installing the windows and casings.

Steve put up a scaffold to get at the roof. He made a series of two-sided, right-angle frames out of two-by-fours with room to slide a diagonal supporting plank into the apex of the right angle to hold the frames against the house. One side lay against the house, while the other, at a right angle to it, formed the horizontal support for the platform.

He covered the rafters with five-eighths-inch plywood for added strength as compared with half-inch CDX. We had decided on the metal roof often seen in Vermont for two main reasons: because for all practical purposes, a metal roof is a permanent installation almost maintenance-free; and because such a roof sheds ice and snow more efficiently than other types of roof of comparable pitch. It appears that the heat in the house warms the metal from below, and the ice and snow slide off comparatively readily without completely thawing. The cost is comparable to asphalt shingles. The metal sheets for our roof cost $226.25.

As Steve hammered away on the roof, Walt set himself up outside with sawhorses, power saw, tape, hammer, and nails and went to work on the plywood sheathing, using half-inch CDX. He first laid vertical sheets all around, fitting them around the window and door openings and nailing them to the studs. He got pretty good at it. Since our walls were 11′1″ high, he covered the top portion with sheets lying horizontally across the tops of the vertical sheets below. The sheets on top, of course, had to be trimmed lengthwise to fit.

I followed Walt around, installing the windows. I had heard that the exterior casing of windows and doors should wait until the siding has been applied over the sheathing, so that the siding can be brought up flush with the edges of

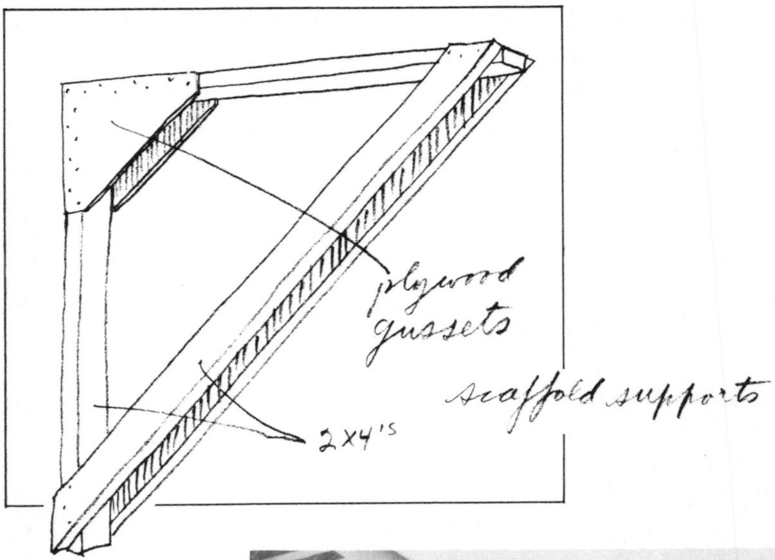

Steve's ingenious scaffold. Scaffold supports are nailed to the building and braced with two-by-fours.

the plywood around the windows and doors, and the casing laid over the siding, so that the casing stands out from the side of the house. Nevertheless, I tried it both ways, impetuously casing several windows without waiting for the siding, later simply butting the siding up against the casings. They look all right to me. The others I cased over the siding.

We used Steve's ingenious scaffold to get at the upper half of the south end with the sheathing and siding, and even with a stepladder on top of the scaffold, it was something of a stretch when we got up toward the ridgepole.

Sheathing completed, metal roof in place, and weatherboard siding up on the south wall.

On the north end we simply used the roof of the garage, plus a stepladder, to reach our work. Here again we had to cut the plywood sheathing and the siding to conform to the garage-roof angle on the bottom and the house-roof angle on top. Walt was the expert in this department.

During all of this, I thought of almost nothing but the weather. Wooden building materials will take quite a beating from the weather, but enough is enough. It rained almost constantly, it seemed. The plywood of our subfloor was so saturated it began to blister and chip off when we scuffed it with our feet. We kept telling ourselves a little damage to the plywood wouldn't be fatal since no one would ever see it once our finish floor was on. But I worried nonetheless. The loft floor too was perpetually drenched; we saved it from serious damage by covering it with plastic sheeting when rain threatened. The nailheads rusted almost overnight.

The day actually came when our roof was on and I was able to relax a little about the weather, despite an interminable delay in getting our electric heating equipment.

Nevertheless, we had our roof, the shell was complete, the siding was on, and the windows in. Our building was beginning to take shape. It looked like a house. When we reached this stage, we stood back and surveyed our work from all angles with a rewarding sense of satisfaction. All that remained to close us in completely was hanging the doors.

Being a coward, I maneuvered Steve into hanging the front door, but I watched how he went about it, paying strict attention to the details. I thought it would be educational, and it was.

For doors that open in, the interior casing must be in place first because the hinges are mounted either on the interior casing itself, or mortised into the jamb which the casing overlaps. Because we wanted our Colonial-style hinges to show, we mounted them on the casing.

First the door is shimmed in place in the rough opening so that it sits square in the opening and clears it all around sufficiently so the door will open and shut freely. It helps to have two people on hand for this chore—one to steady the unattached door, the other to shim it in place and mark the position of the hinges on the door and on the casing. I hung the bathroom door without help, which was a decidedly tedious job.

In marking the position of the hinges, the actual holes where the screws will go should be pinpointed with a pencil. In fact, having done this with the bathroom door, I dispensed with an actual outline of the hinge itself. The position of the screw holes is the crucial factor.

The door is then removed from the opening, and the hinges are screwed to the casing. The door is repositioned in the opening and screwed in place. If all goes well, it will swing freely. The door should just clear the opening. Any excessive gap between the opening and the door will invite cold air drafts.

Once the door is hinged, you shut it flush with the casing on the side where it opens and install your stops on the other side flush up against the door.

A hint to the do-it-yourselfer: I discovered, the hard way, that it is disastrous to mortise your hinges too deep into the door casing or the door itself, particularly if the hinge is set on the inside against the jamb. The mortise should be no deeper than the thickness of the hinge, and the door should be flush with the casing on the side it opens on. It must not be recessed into the opening. If it is, there is a tendency for the inside edges of the door and the casing to hit when the door is opened, which exerts a tremendous strain on the hinges. If this unfortunate situation is permitted to stand, it is only a matter of time before the strain will loosen the hinges. The door must swing freely.

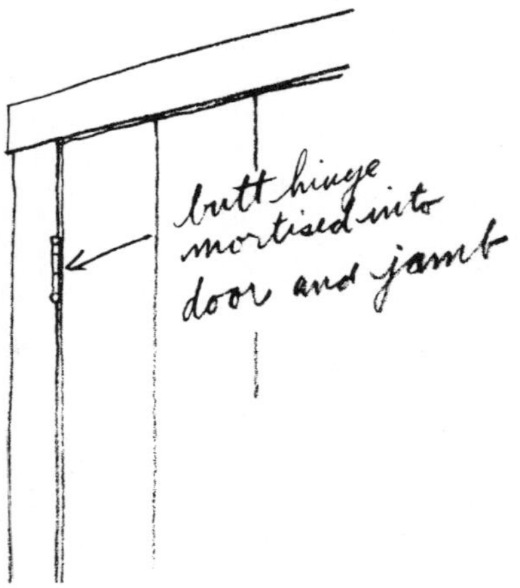

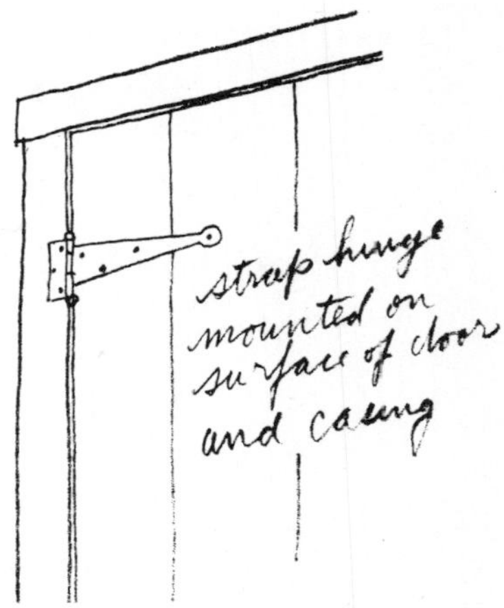

When the stops are in, top and sides, you can install weather stripping all around, even over the threshold, flush with the inner edge of the stops so that the stripping just touches the door when it is closed.

The final step is the handle and latch routine—a fun job if you are patient, but a frustrating nightmare if you're not. The instructions, I think, are often the culprit. I spent almost an entire Sunday installing one of those latch-lock assemblies on the back door (after I had corrected my mortising goof).

"Anyone can install one of these things," I told myself, "by simply following the instructions scrupulously."

Unfortunately, the instructions that come with such intricate hardware items frequently confuse the issue by technical references and omission of details or the inclusion of misleading details and information. Often they come either with sketchy diagrams or none at all. Sometimes they are just incorrect.

My latch-lock instructions said that the lock mechanism would work only if the handle turned in one direction. I

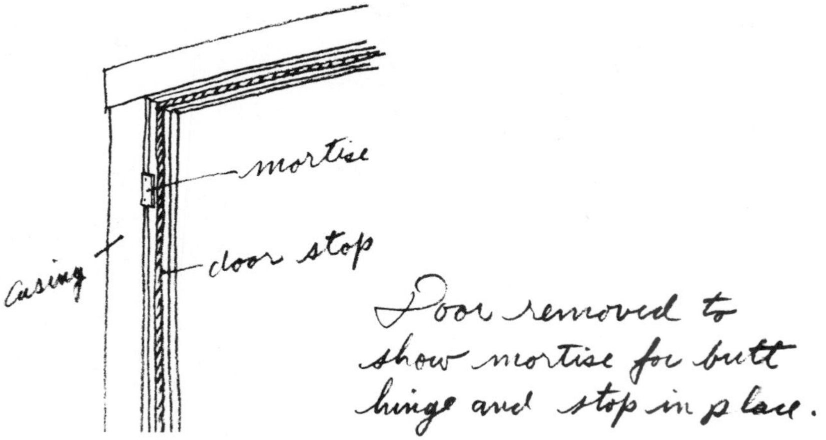

forget whether it was left or right. Everything went smoothly, mortising for the striker, drilling the holes, and so on, until the moment of truth. I tried my damndest to install the gadget their way and squandered three hours in the attempt. I found it was impossible. Their way, the keyhole would have wound up on the inside, making it impossible to open the lock from the outside.

After turning the unwieldy thing over and over in my hands, vainly trying to keep its three components from falling apart, it finally struck me that the instructions might be wrong. It seemed to me it would work when the knob was turned either way. As I held it, I had my wife turn the inside lock flange. It locked, or seemed to.

Gingerly, I installed it my way, initially putting only one screw on each side to hold it while I tested the mechanism in place. It worked perfectly the way they said it wouldn't, and locked just as it should. Steve is enamoured by the gentle metallic click it makes as the latch engages the striker.

Now we were completely enclosed and out of the rain, although we still lived in the garage.

Chapter VII

We Tackle the Interior

Wiring and plumbing; insulation; finish floor and floor finish; sheetrock walls and ceilings; heating equipment; stoves and stove pipes; belated discovery of the importance of a fume vent.

Steve took on the wiring and plumbing, and Walt went ahead with the insulation. Standard 3-1/2″ x 15″ foil-backed fibreglass insulation was stapled between the joists with the foil on the inside toward the living area. This size insulation is made to fit neatly between joists spaced 16 inches apart, center to center, but must be cut to fit odd-size wall spaces. Over the insulation, we stapled plastic sheeting to serve as a moisture barrier to discourage the escape of moisture from the air inside. Total cost of insulation: $159.27.

We laid our finish floor of wideboard Maine pine at about twenty cents a board foot, a substantial savings on flooring material. We considered a hardwood floor but eventually settled for pine because of the savings involved—and the experience of a friend of Steve's who put down a pine floor and found it entirely satisfactory. My wife felt she could do without the drudgery of cleaning and polishing a hardwood floor, and we both felt a modest pine floor was maybe more in keeping with the rustic decor of our mansion.

Our floor is raised three-quarters of an inch above the

plywood subfloor. It rests on strips of strapping spaced "16 on center" and nailed to the plywood floor below. I covered the floor first with a clear wood preservative, then a maple stain which exposes the grain and harmonizes nicely with the gray earth tone of the barn beams. Finally, we applied a coat of clear, non-gloss polyurethane floor coating, the best thing there is for floor finishing, according to many advocates. It dries clear, evenly, and hard. Water spilt on it beads up like water on a cabbage leaf. The coating, applied thoroughly and evenly, appears almost 100 percent waterproof. It is recommended that additional coats be applied periodically—perhaps every couple of years—to reinforce and maintain the hard protective finish, particularly in heavy traffic areas.

The finish floor cost us $169.84 for the lumber, $13.81 for the polyurethane coating, $7.29 for the stain, and $20 for the wood preservative—or, say, $200.

We bought enough preservative to coat the loft floor and the staircase, which Steve contrived out of two-by-tens, again with Euclid's assistance. So the full cost of the preservative cannot be attributed to the downstairs floor.

One of the picayune assignments in laying a floor is countersinking the nails, both to hold the floor down firmly and to get the nails out of the way of pedestrians' feet. A friend of ours sat on his haunches for two afternoons pounding away with hammer and countersink. He said it is a job for a moron, necessary though it is. By the way, proper flooring nails are threaded, on the theory that they hold

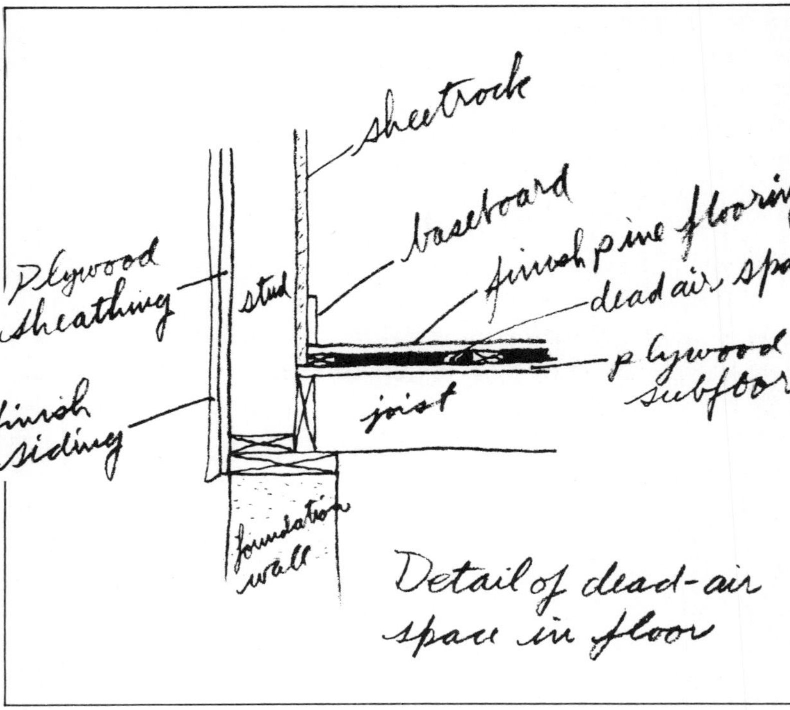

more firmly and have less of a tendency to pop up under the constant movement of weight above.

"There ought to be a law against sheetrock," one of our suppliers commented one day as I was ordering $163 worth. I quickly learned what he meant. Like plywood, the stuff comes in four-by-eight-foot sheets and, like the plywood on the outside wall, is nailed over the studs on the inside (theoretically two sheets to each eight-foot section) to form the interior walls. But, again, much cutting of the material is required to fit the odd-size spaces and angles at each end of the house upstairs.

Cutting is done with a knife and a straight edge—we used

a level. The sheets must be scored through the paper facing on one side, then snapped and broken along the line. The paper facing on the other side is then cut along the break. The sheets are extremely heavy and must be handled gingerly to prevent inadvertent damage to the plaster core.

Cut-outs for such things as wall receptacles (for light switches, etc.) must be tackled with drill and saw. The by-product of all this is a miserable, fine, gray dust that defies every known method of housekeeping.

Most of our cutting was done on the garage floor. Clouds of dust lay suspended in the air and eventually settled on everything. Coarser dust piled up all over and was constantly tracked over our newly finished floor. We swept out, if not every evening, fairly faithfully at frequent intervals, but always there was a residue adhering to feet and floor.

In spite of the dust, sheetrock is the popular method of covering inside walls. It can be painted, wallpapered, or paneled over, which versatility probably explains why there is no law against it.

A note of caution: although one man can, with a little muscle, install a full sheet on a vertical wall where there is a base to rest it on, putting up the overhead sheets on a ceiling is a job for three men. The stuff is so heavy we found the job almost impossible for two. You need two men to hold a sheet up at each end and a third to tack it in place. One man can then finish the chore.

As with the application of plywood, it is a good idea to indicate the position of the studs or joists with chalk marks on a beam or stud above and on the floor below before nailing your sheet up. This makes it much easier to locate and mark off your nailing lines. In fact, the same general principles apply to installation of plywood and sheetrock, the chief difference being that one is on the outside and the other is on the inside.

The seams where the sheetrock sheets meet on the studs are sealed with plastering compound and perforated tape. The compound is applied with a putty knife along the seam in a thin, even coat. It should be worked into the seams and cover the width of the tape. The tape is then stripped over the compound bed, and another layer of compound applied over the tape. The nail heads also should be covered with the compound, which means that the nails must be driven deep enough so they don't protrude beyond the surface of the sheetrock. Otherwise you wind up with a sloppy-looking job, and the protruding nail heads will deface your final finish, wallpaper or paint.

For the moment, our plastering job remains incomplete, but it is not a priority item and has little to do with the insulation factor. Time ran out. Steve returned to school. Walt went to work for an oil-heating contractor in St. Johnsbury.

Some time in early August we had ordered our electric heating equipment through a friend of Steve's with legitimate credentials as a contractor. It was one of those "I can get it for you wholesale" deals, promising to save us $100 or so.

We waited five or six weeks, and found the order had somehow miscarried and the equipment hadn't yet been shipped from the supplier in Rutland. Another four or five weeks went by, and we still had no baseboard heaters. We were assured they were on the way. But we were, by now, getting into October. The temperature had dropped to freezing a few times and my wife, who had gone to New York on a family matter, said she would not move in until there was heat.

Ultimately, we gave up on the big deal. I never actually learned what went wrong, but we did not get the stuff. Steve and I went to Burlington one day in the pouring rain and bought the equipment we needed. We did not pose as contractors. Steve simply told the wholesaler that I was paying the bills and he was doing the job.

We were given not only the contractor's discount, but a quantity price discount as well. From what I gathered from the price list I scanned briefly, our discount was substantial, amounting to almost a third of the list price, retail.

I wrote the man a check for $270, and we drove home with the equipment. Steve installed it in two or three days.

Meanwhile, during the marathon wait for the heating equipment, we put in the wood stoves. Again, Steve called on Euclid to cut the tin sleeves that line the holes in the roof and the egg-shaped holes in the flashing we fitted around the stovepipe where it emerged from the roof. This is a fairly complex proceeding, requiring the use of geometry to calculate the cuts so they fit snugly. If the fit is poor, particularly that of the two overlapping sheets of flashing around the pipe, you invite leaks. Cutting holes in your nice leak-proof roof is an anxiety-producing experience in the first place, but if the seal is snug and plastered with enough roofing cement in the right places, it won't leak. Ours didn't, at first. But winter has taken its toll. The snow and ice sliding off the metal roof, alas, have knocked the pipes askew and opened up the seals. The damn things leak and, at 14 below zero, it is a shade too cold to get up there and tinker with them. But I will think of something.

After the cuts in the flashing are calculated, you make your initial cuts smaller than the calculated dimension. An inch and a half is good. Then a series of cuts are made radiating out from the perimeter of the arc so that the tabs thus formed can be bent up to fit around the pipe. The top sheet of flashing of course overlaps the bottom sheet so the water has no opportunity to run under the bottom sheet. The upper edge of the top sheet must be sealed thoroughly with roofing cement for the same reason. I also daubed roofing cement over all the nail heads. A standard-size cap made for the purpose is fitted on top of the pipe to keep water out.

I made the mistake of smearing too much roofing cement at the base of those tabs where they touch the pipe. We found that when the pipes heat up, a horrible black sludge, presumably an amalgam of melted tar and water, runs down the outside of the pipe onto the floor. I had to string up foil pie pans under the elbows where the vertical pipes join the pipes running horizontally to catch the sludge. It is a nuisance. Mysteriously, sludge sometimes runs down the pipe in a thinner consistency even when there is no fire in the stove. Fortunately, roof cement is noncombustible, but the sludge produces a depressing mess. It splatters. As I said, I'll have to think of something.

We installed the cook stove in the kitchen area, running a pipe vertically from the stove, then horizontally for a few feet, and finally a second vertical run taking the pipe through the roof. I lined up the parlor stove in the living room with the box stove in the garage and connected the two with a long horizontal run under the garage and loft ceilings, through the back wall of the garage. The vertical pipe running up through the roof is attached to the horizontal stretch with a T connection.

One of the theories of stove installation is that the more length of pipe you have inside, where the air is warmer, the better. First, you get more heat, and second, your draft is likely to be better. The axiom is that hot air rises.

We discovered, however, that although our stoves draw beautifully in still air or even in wind from a certain direction, a down-draft is disastrous. It blows the smoke back into the house. I intend to correct this flaw, maybe in the spring, by rigging up two horizontal sheets of metal where the smoke emerges from the pipe so that the wind will have to pass over the pipe horizontally and cannot blow down into it, theoretically at least. This device was suggested by a friend, and it seems to me it just might work. As can be seen, our difficul-

WE TACKLE THE INTERIOR

Stephen T. Whitney

Kitchen stove pipe in place, with cap—some problems yet to be resolved.

Stephen T. Whitney

Kitchen-dining area seen from living room. Note wide-board pine floor. Framed painting of floor plan hanging on wall to left was done by Walter Rae's 13-year-old niece, Susan, in honor of the project.

ties with the stoves are far from over.

Toward the end of the building season, late August and September, we ran into out-of-stock problems at the lumber yards, another good reason for planning as far ahead as possible and getting your hands on needed supplies early. Fortunately we had almost all of our supplies by this time, but there are always last-minute pick-up demands.

You may run short. We found we needed four sheets of plywood quite late in our building program to floor over the south end of the house we had left open to give access to the plumbing and wiring networks. The lumber yards, it turned out, had exhausted their supplies. Finally I phoned a supplier who had some left—and I discovered why. His prices were exorbitant. But we paid his price because we had to have the plywood. The lot fell to my wife to drive the 40 miles to get it.

In September after our shell was enclosed and I started work on the inside trim, I could not buy a stick of one-by-two, one-by-four, and other sizes of No. 2 pine. At one point all I could get was one-by-twelves and I had to rip them down to the sizes I needed. I ripped literally hundreds of feet of wood. With a good blade, it isn't an impossible task, but it is an added chore. Also you lose the planed edge on at least one side. You'd need a jointer to plane it smooth.

Don't forget the stink pipe, or fume vent, as it is more conventionally known among genteel folk. No doubt you have seen them jutting mysteriously from the rooftops of the land, and even may have wondered what they were. Their purpose is to carry the noxious fumes generated by bacterial action in septic systems up into the yonder reaches, mercifully bypassing the indoors, so that they don't offend the olfactory senses of the occupants. (And also to allow escape of the air displaced by water when a toilet is flushed or water is let out down a drain. Ed.)

Precisely, the vent is a pipe, in our case a one-and-one-half-inch (inside measurement) PVC plastic pipe, leading from the top of the main waste disposal line where it intercepts gases emanating from the septic system beyond. The pipe runs up through the house, usually in an out-of-the-way corner, and out through the roof.

If you overlook this little refinement, your neglect surely will one day betray you and your loved ones in a harvest of remarkable unpleasantness, staggering to the nostrils, if you are disposing of household wastes via the septic tank method. Oh, you may get away with it, as we did, for more than a year, but your day will come. Ours did after a two-week siege of 90-degree weather which obviously inspired vigorous bacterial action in the malicious, somnolent monster—the septic tank.

It is not that we failed to anticipate the problem; simply my sloth in following through. Steve installed the line, running it up through the kitchen floor in the corner, and stopped there. He capped the pipe, and left the rest to me. The chore was relatively simple, involving merely cutting holes in the loft floor and the roof and connecting a length of plastic pipe to the upper end of the pipe in the kitchen. The pipes slip neatly together inside a connecting sleeve. The opening in the roof, of course, must be sealed with flashing and roofing cement, but the job entails no fire hazard and so does not require the extra care demanded in installing a chimney pipe.

But I put it off, basking in the sublime sense of fulfillment derived from finishing the house. In the back of my mind I knew it was a job that had to be done, but as long as the septic tank cooperated, I was in no mood to hasten the trauma of hacking a hole in a perfectly sound, watertight roof.

A year passed. And then two weeks of tropical heat in

August 1973 stirred a frenzy of ferment in the potent septic tank brew, choking the house with an overpowering fragrance.

"All this rain," I said to Isabel, "has gotten to the cellar. It's the musty smell of the cellar, maybe. I'll open the vents in the foundation and air it out. Maybe that will do it."

Isabel seemed skeptical, perhaps because my evasion lacked the ring of conviction.

Airing out the cellar did not do it, and presently I acknowledged what I knew from the outset, the septic source of the problem. Isabel appeared relieved.

"That's what it smells like to me," she said, thankful for my acceptance of reality.

I installed the pipe forthwith, to the intense relief of our olfactory senses.

The pipe bows slightly as a result of imperfect alignment of the holes in the ceiling and roof, but it does the trick. The malevolent fumes have been vanquished. The job took me about three hours. Cash outlay: zero, since we had the pipe, the roofing cement, and a scrap of flashing left over from the materials bought in the summer of building. Miraculously, the roof is as watertight as ever.

Chapter VIII

Stephen T. Whitney

Author Walter Rae.

What We Spent and How We Spent It

Cost breakdown—over-all and month by month; labor; "special" expenses; artesian well and septic system; hidden extras; estimating costs in advance and certain intangible benefits.

As I said in the introduction to this book, I know what we spent almost to the dollar and, for practical purposes, what for. Of the initial (see p. 26) total of \$7,820.77, \$2,321.50 went for wages* and \$2,524.50 for lumber, the major items of building costs in the economy-style approach as well as in the conventional way of going about it. The job could be done for more, obviously, or less, depending on materials, climate, and the size of your house.

The following breakdown shows at a glance where the lion's share of our outlay went:

Lumber			\$2,524.50
Wages*	Steve	\$1,456.00	
	Walt Jr.	865.50	
	Total		2,321.50

---

*A do-it-*all*-by himselfer working weekends and evenings could discount this item.

Electric supplies, including baseboard heat	630.84
Concrete pouring	314.25
Metal roofing	231.25
Sheetrock	213.07
Plumbing supplies	202.17
Insulation	159.27
Windows	132.77
Doors, storms	115.20
Stovepipe	63.44
Total	$6,908.26

If my addition is correct, this leaves $912.51 of the $7,820.77 unaccounted for. It went—like water through a sieve—for innumerable miscellaneous items such as weather stripping, plastic sheeting, nails, hardware, plastering compound, stain, wood preservative, polyurethane floor covering, and so on.

For example, there are about $40 worth of nails in the house, and $37.85 in stain, wood preservative and polyurethane floor covering.

From time to time as we went along, I jotted down total expenses to date. The figures do not indicate progress, as they often include materials bought far in advance, but they may be of interest anyway if for no other reason than to show the relentless outgo of funds in even such a modest undertaking as ours:

June 5	$ 32.00
June 15	381.57
June 26	815.05

August 4	1,548.74
August 10	2,330.49
August 21	3,118.85
September 11	5,635.79
September 14	5,844.11
September 19	6,081.91
October 4	6,744.93
October 14	7,160.86
November 24	7,744.06

We reached our final total in December with the installation of the electric range and some lumber I bought for inside trim. I don't know why I omitted entries in July—too busy maybe, or too harassed.

As everyone knows, labor is a major factor in construction costs. I was fortunate in having the services of my two sons, Steve and Walt, Jr., at what I concede were slave wages. (Obviously, though, anyone who has the time and knowledge to build his house all on his own on weekends and evenings, say, would not have this particular expense.)

We paid Steve $3 an hour—not our idea, his. At the outset, somewhat hesitantly we offered him $4, and said the figure was negotiable. But he balked, saying his objective wasn't to get rich.

Walt, who had no experience and knew almost nothing about the use of tools, was paid $2 an hour, later boosted to $2.25. It turned out to be a bargain. Walt is a willing worker, strong as a bull, and learns very fast. He applied most of the exterior plywood sheathing, practically all of the weatherboard siding and insulation, and much of the sheetrock on the inside walls and ceilings.

I learned from the experience that untutored labor of the

right sort can be invaluable. The individual tackling such a project without the advantage we had in the persons of my two able sons is likely to run into substantially higher labor costs, unless, of course, he undertakes the job on his own and extends his timetable.

Nevertheless, one doesn't always have to pay the going rate—not even for labor. Last year, a summer resident neighbor was able through connections in town to hire a seasoned carpenter for $5 an hour—which was well under the going rate—for some work he wanted done. The cut-rate carpenter turned out excellent work, the neighbor said. So the home builder lacking the plus of a resourceful son or two can still get help at fairly reasonable rates if he inquires around, perhaps one experienced carpenter like my neighbor's and an inexperienced helper. He will find this cheaper than a contractor.

My record of our costs was intended to reflect only those expenses directly applicable to the actual building of the house. I included the electric connections installed by the power company since almost always expenses of this kind are encountered. But I did not include appliances and furnishings since these items are optional (more or less, as I found out), depending on taste and the degree of one's reliance on modern conveniences.

If one is content, as most are not, to heat water on a cook stove, one doesn't need a hot water heater. Such, in fact, was my original intent, but I doubt that it was my wife's. (Concession No. 2: we bought a hot water heater, currently out of order.)

I should mention that we made two decisions bearing on costs that a contractor would ignore, unless you or your architect specified them, because neither is standard practice in home construction. We decided to use two-by-six-inch wall studs instead of two-by-fours or regular stud timbers which are even smaller. Two-by-fours, by the way, have shrunk to 1-1/2″ x 3-1/2″ over the years (see Appendix B). But two-by-fours and the smaller stud timbers are the general rule, even in this Vermont climate of severe winters. Our idea was to increase the dead air space in the walls between the insulation and the outer plywood shell for better insulation.

For the same purpose, we installed a double floor—the half-inch plywood false floor laid over the floor joists was covered with a finish floor of wide board Maine pine. Pine strapping (1″ x 2-3/8″) separates the false and finish floors, leaving a three-quarter-inch dead air space between the two to impede the intrusion of cold air from below. Incidentally, rough-cut strapping is about half the price of comparable (say one-by-two) No. 2 planed pine, and just as suitable for use in inconspicuous spots. I even used some for exterior molding under the eaves and for inside window casing, not being a stickler for fancy finish.

These "extras" cost possibly $250 more each, but the added cost is doubtless justified in terms of winter comfort in subzero temperatures.

The cost of getting water and installing a septic system of course cannot be excluded. If you are not lucky enough to have a never-failing spring, you must "go" for water—and what is a home without a nice antiseptic bathroom?

We had an artesian well dug by a local man at a basic cost of $7 a foot, six-inch casing included. The immersible pump and storage tank were extra. The cost: $978.01, including the extras, plus $13 for electric hookup. The basic charge was considerably better than the going rate of around $10 a foot. We breathed easier when it was over because the cost of digging an artesian well is anybody's guess. It depends on how deep you must go. Our man hit a good flow of water at 80 feet, but some wells in the area are three times that depth and outlays of $2,000 to $3,000 are not unusual.

Our septic system, including the 1,000-gallon tank, pipes, and leach field, came to $384, probably less than we could have done it for ourselves because of the labor involved. A local man did the honors here too with a backhoe.

Our grand total to date then is $7,720.77, plus $978.01 for water and $384 for septic system, or $9,182.78—well under $10,000 for a solid, two-story house with the essential services. It is even reasonably weather-tight despite that darned draft coming in around the front door.

Except for the window and door casings, the house lacks interior finish, but there is no hurry now. The blitzkrieg is over.

You cannot possibly anticipate all the hidden expenses. There are many, and they add up. For example, you might find you needed a water softener. There is no way of knowing in advance how much tapping a water supply is going to cost (if you are locating beyond the reach of a public water supply). And what if you should strike ledge when you are digging your foundation? You would have to blast.

Again, if you plan on any inside finish work by a professional cabinetmaker, you may find the cost is higher than you'd expect. We did. Like $600 to $700 for kitchen cabinets, estimated, compared with $300 we had allotted. We found we had to put a butane gas heater in the garage to make it comfortable as a shop or study when we had expected, unrealistically perhaps, to heat it with the box stove: another $132. And so it goes. The expenses mount for necessities required for comfort and convenience and, in our case, continue to crop up.

My first thought after glancing at my expense records and making preliminary calculations was: well, I can tell the folks we did it all for under $10,000, including inside finish such as the kitchen cabinets, which I was not in the mood to tackle when the time came. We didn't. To be honest, I'm going to hike that estimate to $11,000 or maybe $11,500 for the complete job. We still have a bargain.

I would recommend to other would-be home builders not to play it too close financially. On the basis of our experience, anyone planning to take the plunge would do well to carefully calculate his potential expense, including everything he can think of, and then add $2,000 or, better, $3,000 for unforeseeable contingencies. Avoid the trap of setting your sights to the last penny of your reserve. You will regret it. For example, if your available funds amount to $12,000, I'd plan a building program of no more than $9,000. If you can lay your hands on $15,000, I'd keep the building cost estimate to $12,000. Similarly, if you have $18,000, you'd better figure on $15,000 for the complete job.

It is sometimes said that, if a man takes six months off to build his house, he must, to be honest, charge his "lost" salary to his costs. On this basis, a man who earns $12,000 loses $6,000. In our case, this would up our total costs to about the figure we would have spent if the job had been done by a contractor. To me, this is a kind of ephemeral question that depends on individual circumstances such as: was the home builder employed at the time of his endeavor? In some cases, the man who tackles such a project is in a financial position to do so with a little surplus at hand to take up the slack. That is why he has the choice. A venture like ours often takes place during a period of transition in the lives of its perpetrators, when one style of living is being exchanged for another, so that there is a natural break or diminution in income. I will leave it to others to decide for themselves.

But one who allows this consideration to deter him cheats himself of what Burroughs called one of life's greatest pleasures and the opportunity to muse "in sweet reverie upon the naked joist." There is no dollar value that can be placed on these intangibles. And to the man bent on such an undertaking, they may very well be the highest return he looks for from his investment.

Possibly, in a small way, the whole issue is akin to the question of why men go to the Poles in dog sleds, or climb Mt. Everest; there is no money in this sort of achievement, but there *is* an enormous amount of satisfaction and pride of accomplishment.

Reading nook under the stairs that lead to the loft. ▶

WHAT WE SPENT AND HOW WE SPENT IT

Stephen T. Whitney

Chapter IX

We Cut Some Corners

Buying windows and doors; plumbing fixtures, new versus used; weatherboard and cordwood—you can get it for less—look around!

What is true of the labor market is also true elsewhere. It is folly to pay top dollar for everything. You don't have to.

Take windows. Before we left New York, we made several pilgrimages to the carnival-like Four Acres of Junk in Stamford, Connecticut, purveyors of second-hand by-products of the house-wrecking business. There we bought almost all our windows, except storms, for $10 to $15 apiece, including a fine French casement window complete with screen and crank-action hardware for opening the sashes for $15. We also bought a splendid, large four-light window with screen and storm attachment for $15. Our acquisitions included two new, smaller windows of identical size with screen for $2 each. Apparently they were discontinued items. I installed them in the kitchen—one as the permanent window, the other as the storm, clipped in and easily removable.

The two $15 jobs were a bit battered and the tracks of the storm-screen attachment of one were clogged with grime. Otherwise they were intact and perfectly serviceable. All they needed was cleaning and minor restoration work such as straightening here and there.

We also salvaged several extraneous matching storm windows from our house in New City and paired them up in the Vermont house as combination permanent installations and clip-in storms.

In short, of the ten windows we installed, three, including the $4 pair, are new; the rest are second-hand. We bought other second-hand windows in Vermont to serve as storms, matching them up with the windows we had as best we could and doing some fancy fitting on the inside frames to make them fit. Our total investment in windows was $132.77. If we had bought them all new, the cost could easily have exceeded $1,000.

By the way, all my storms clip in from the inside, fitting flush into the interior window casings. It seems to me that this is the simplest arrangement both for convenience and ease of installation.

Our doors and storm doors, with the exception of the bathroom door, we bought new, largely because no "quaint" old doors showed up. You can invest only so much time in bargain hunting for charming appurtenances without running the risk of having no door at all by the time winter sets in. The bathroom door cost $10.

As for such fixtures as sinks, lavatories, tubs, and toilets, you can whittle down the cost dramatically by buying used, even antique, merchandise. But in buying items that must be hooked into your plumbing system, be sure that the fittings come along with them whenever possible. You will save lots of time and avoid the frustration of searching for obsolete

Stephen T. Whitney

This antique soapstone sink cost $25.00, including yellow paint.

fittings. You will have problems enough as it is.

For example, we bought a beautiful antique soapstone sink (smeared with the inevitable yellow paint) for $25. The drain, which happily proved watertight, was intact. A short length of pipe protruded from under the drain, threaded at the end where it was to attach to whatever piping was used. But a coupling was missing. I spent the better part of two days looking for it in plumbing supply outlets only to find that what I needed apparently was an Essex coupling which hasn't been manufactured in 40 years.

As it turned out, Steve was able to improvise the connection with a combination of PVC plastic pipe fittings. Not every plumber is as ingenious.

Our tub, the rather handsome cast-iron variety sitting on feet, and toilet we bought for $20 from a man who was remodeling his bathroom. Most of these items we spotted in the merchandise-for-sale column in the local newspaper or heard about via a trading-post program on our local radio station. Our lavatory, with all the fittings, cost us $10 at a roadside junk shop.

In other words, all the fixtures that tie into our plumbing system (except the new hot water heater) cost us $55. And there is something to be said for the suitability of these fixtures in the basically 19th-century New England setting we more or less simulated in Vermont. They are compatible with the wood stoves and barn beams.

The weatherboard for our siding, 1,900 board feet of it, I got from a supplier in Wolcott, Vermont, whose chief business is processing weatherboard and filling large orders for contractors, wholesale. His only retail business is in "seconds," boards that don't meet standards for top-grade lumber as a result of flaws such as knotholes or split ends. We paid 20 cents a board foot—about half the price of firsts at a lumber company.

A friend who had used some steered me on to the Wolcott deal. He said there would be very little waste and he turned out to be right. We had to discard very few of the boards and found we could salvage many with imperfections by trimming when we needed shorter pieces. Some trimming would have been necessary anyway.

Actually, we wound up with two or three hundred feet left over, currently under the snow—covered, of course.

Total bill for our weatherboard: $391, tax included. This amount is included in my total figure for lumber. Almost any other acceptable material for siding a house the size of ours would be double that, and more.

Although somewhat remote from the business of building a house, the price of cordwood is another example of my contention that you can often get it for less. Judging by the ads in the local papers, firewood goes for $50, even $60, a cord in this part of the country. But if you carry your investigation a little further, you can do better.

When I commented on these figures, with an ulterior purpose uppermost in mind, to the man who dug our well, the well-digger merely laughed—and immediately came up with the information I wanted.

"You don't have to pay that for wood. There's a fella right down the road here who'll sell it to you for $25 a cord. Good wood too. He don't believe in them kind of prices, I guess."

He told me how to get there and gave me the man's name.

"Can I use your name?" I asked the well-digger.

"Hell, yes," said the well-digger. "Hell, yes."

Chapter X

We Move In

Snug and warm at last; a sense of fulfillment; kinks still to be worked out; electric heat costs; adjustment to rural life (it is different); country pursuits; have we found our Shangri-la? Time will tell . . .

It was October. We turned on the heat and moved in.

My wife, who has an instinct for homemaking, transferred a house into a home in a single day of whirlwind activity. Under her direction and watchful eye, our gear, or most of it, was moved into the house. Isabel has a knack for putting things in the right place, and by night the transformation was just about complete. Familiar things were in place. Even the plants looked at home. We were snug and warm despite the wind and the first snow flurries outside.

With a profound sense of fulfillment, I sat by the fire in the parlor stove in the house I had created myself, my mark upon the land that will survive me by 200 years or more, if nobody lays a finger to it. I was pleased and comforted by what I saw. The color scheme of the room in the fading daylight was warm, despite its unfinished rawness. Everywhere there was the soft presence of natural wood, with its lovely, shifting patterns of grain—the floor, the staircase, the beams, the window trim. The gray sheen of the stately, ancient barn beams, with their craggy, time-etched countenance, was soothing. They spoke to me. As I looked at them, those venerable witnesses of life and death, they seemed to return my gaze with a suggestion of gratitude. Why not? I had, after all, rescued them from the undergrowth where they had fallen and restored them to a position of dignity. They were here before me and will remain after I am gone.

The results, wherever I looked, were highly satisfying in spite of the black sludge running down the stove pipes, trouble with the water heater, disappointment at the need to limit the use of the wood stoves, the expense of electric heat, the draft around the front door, and other assorted frustrations.

There are all kinds of kinks to be ironed out. We just discovered, for example, that our water is too hard to make an electric water heater practical, at least without a water softener. As it is, heavy deposits of calcium and magnesium corrode the heating element very rapidly, putting it out of commission. The element would need frequent replacement without a softener system. We could switch to an oil- or gas-fired heater, but these alternatives present problems since we have neither oil nor gas, except for the butane heater in the garage. All three solutions are expensive. Meanwhile, the solenoid on the water pump has taken to sticking. It will have to be fixed.

As for the cost of electric heat, the local utility company offers a combined rate in the case of homes heated by electricity. The rate is just under three cents per kilowatt hour for the first 600 hours, which the average homeowner would normally use for light and appliances. Thereafter, the rate drops to 1-1/2 cents per kilowatt hour. Our latest bill for 1,880 kwh was $40.26, plus $1.21 state sales tax. It represents mainly our use in November, when temperatures here were above normal. But our meter indicates that we are now using about 3,500 kwh a month, so the December bill should run between $60 and $70. My guess is $63 or $64. The bill may be even higher in January and February, the real winter months in northern Vermont.

For the time being, we are surrounded by bare sheetrock, unpainted woodwork, and exposed, unplated electric switches. Plastic sheeting forms the ceiling over our living room where the room is open to the loft ceiling. I tacked up the drab sheeting one day when the temperature hovered around zero to keep some of the heat downstairs.

Our flight to the woods is complete as far as the basic housing need goes. The period of adjustment continues. We are a long way from our two daughters and most of our friends. Rural living is new to Isabel. It takes time to get the hang of it. We are only about six and a half miles from the state capital, Montpelier, but we are surrounded by open fields and trees instead of by the familiar brick walls, asphalt, city lights, and whining sirens. A dirt road stretches some two and a half miles on one side and a mile or so on the other. The countryside is magnificent. This is something that we do not yet take for granted—a beauty that seems to renew itself each day in the changing patterns of daylight and seasons to present a fresh countenance. No traffic hums outside. There is silence, broken now and then only by a dog barking in the distance. It is peaceful, and we breathe cleaner air.

Still, pastoral charm aside, living in the country, for those who are new to it, takes getting used to. At first we both had a definite feeling of isolation. Our oldest daughter—who did not marry the farmer's son down the road—felt cut off from us. (Concession No. 3: we put in a telephone.)

I think, though, that the feeling of isolation (which beset even Thoreau—for an hour—when he first went to Walden) retreats as the transplanted urbanite or suburbanite cultivates

Stephen T. Whitney

The staircase and barn beams as seen looking towards the kitchen.

an appreciation for such blessings of rural living as privacy and quiet and space to move on the land—north, south, east or west—and finds new rewards in the indigenous pursuits of country folk. Most city folk find it very hard to sit quietly for any length of time without a sense of panic, inured as they are to the compulsive pace of modern society and dependent on external stimulants to ward off boredom. The ability to derive pleasure from peace and simple surroundings must be learned like other arts.

Canning, cutting wood, and growing vegetables are productive and useful pursuits, economically expedient and esthetically gratifying. They promote the same sense of independence derived from building one's own house. If I cut my own wood, I don't have to buy it. Not only that, but I recognize individual pieces as I feed them into the fire. I am closer to the realities of wood and fire. If I grow my own vegetables, I am less dependent on the supermarket. I am helping to sustain myself in more ways, and in a more real sense than simply paying out dollars for the produce of others. Fresh vegetables from your garden and vegetables bought in the markets are two different things. Isabel recently paid 60 cents for three tomatoes—20 cents each! They were tasteless.

I doubt that loneliness will preoccupy my psyche when I am scratching in the earth to grow my own vegetables. Isabel may find a comparable liberation in putting up a few jars of tomatoes. That is the way of country living, but one cannot expect the conversion to be instantaneous if one has always depended on Del Monte and Campbell cans.

Then there is the salient question of the propriety, in a world ridden by violence, suffering, and ugliness, of retreating to the comparatively idyllic environment of the quiet countryside, remote from the massive social struggle. Some purists are bound to call it escapism. It is not. Having the

Stephen T. Whitney

The living area with a beautiful, big, black parlor stove (not yet connected) and an old-time railroad-station calendar clock. Inside storms clipped in.

choice, I have come here to cultivate a more satisfying life-style in closer contact with nature than is possible in the megalopolis.

The fact is that few of us, no matter where we live or what we do for a living, dedicate ourselves to social reform or the amelioration of suffering. Rather we are a culture committed to material gain and free enterprise. For most of us, it is a matter of climbing to the top of the salary scale as fast as possible just to keep up with the mortgage and car payments. So the question, for practical purposes, may even be irrelevant. Helen and Scott Nearing did not think so. They considered it an ethical issue, going to the roots of the social order. Their position was that each of us, who has the choice, has a responsibility to live as well and as fully as his intelligence and talents permit.

It can also be argued, as they felt, that if congestion is one of the social evils of the decaying urban centers, one's very presence there contributes to the evil. Perhaps, after all, I do serve a social purpose by merely removing myself from the urban throngs. And it may be that, by doing so, I set a tiny precedent for others to follow, if they wish, in the hope of creating a more rational social order. As I see it, it would be a culture based on closer contact with nature and individual productivity in a range of activities rather than specialization and the mass-production techniques of modern urban-industrial society, where an individual's entire working life, and income, may depend on one specific operation, such as setting a bolt, on a production line—endlessly repeated, day after day, year after year. Doctors, engineers, and educators are specialists. The monetary rewards are greater. It has gone so far that some men have succumbed to the age of specialization to the point where you hear them confess proudly in apparent defense of their specialty, "Why, I couldn't even drive a nail." What they seem to be saying is:

"Acknowledge my authority in my particular field, but ask no more of me." They make a virtue of it since men have a need to believe in their way of life.

The prevailing social order, it seems to me, erodes individual initiative and relegates nature and the manual skills to virtual obscurity, thereby diminishing man's dignity and self-reliance. I chose the country. To my mind it provides the climate that gives me the best potential for becoming my own man. Our hope is that, having made the choice, it is the best one for us.

Thus far, we have found living costs here just about what they were in suburbia. Property taxes may be a little lower, but recently there have been some rude surprises for many property owners. The big developers are moving in, forcing taxes up as municipal services mount, and the farmers are being forced to sell out because the tax rates are outstripping their return from the land.

About the only regular expenses we have eliminated up to now are the monthly bills for water and garbage collection. I am my own refuse man. But our water system costs are adding up to the equivalent of a fair number of monthly bills.

In the summer, when the frost is gone, I plan to cultivate a modest garden by the organic method. If successful, it may have some impact on the grocery bill as well as our adjustment to the mystique of country life.

Whether or not we have found our Shangri-la, time and our capacity to adjust will tell. Right now we are weathering our first Vermont winter with an anxious eye on the water pipes.

Stephen T. Whitney

"Something is there on the Vermont hillside that was not there before . . ."

Some Good Practical Manuals of Home Construction

DWELLING HOUSE CONSTRUCTION by Albert G. H. Dietz. MIT Press.

HOW TO BUILD YOUR HOME IN THE WOODS by Bradford Angier. Hart Publishing Co., New York, 1952.

MODERN CARPENTRY by Willis H. Wagner. Goodheart-Willcox Co. Inc., South Holland, Illinois, 1973.

WOOD-FRAME HOUSE CONSTRUCTION (Agriculture Handbook No. 73) by L.O. Anderson. U.S. Department of Agriculture. U.S. Government Printing Office, Washington, D.C. Revised July, 1970.

BASIC CONSTRUCTION TECHNIQUES FOR HOUSES AND SMALL BUILDINGS SIMPLY EXPLAINED. Prepared by the U.S. Navy (Bureau of Naval Personnel). Dover Publications, Inc., New York. 1972.

THE OLD FARMER'S ALMANAC GUIDE TO LUMBER AND NAILS

LUMBER WIDTHS & THICKNESS in Inches

NOMINAL SIZE	ACTUAL SIZE Dry or Seasoned
1 x 3	¾ x 2½
1 x 4	¾ x 3½
1 x 6	¾ x 5½
1 x 8	¾ x 7¼
1 x 10	¾ x 9¼
1 x 12	¾ x 11¼
2 x 3	1½ x 2½
2 x 4	1½ x 3½
2 x 6	1½ x 5½
2 x 8	1½ x 7¼
2 x 10	1½ x 9¼
2 x 12	1½ x 11¼

NAIL SIZES

The nail on the left is a 5d (penny) finish nail; on the right, a 20d common. The numerals below the nail sizes indicate the approximate number of common nails per pound.

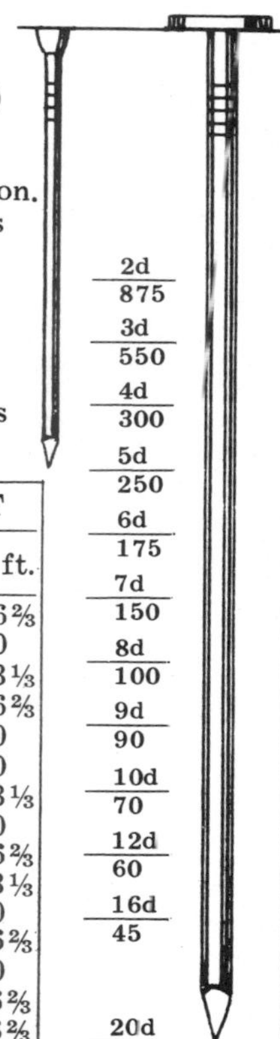

LUMBER MEASURE IN BOARD FEET

LENGTH Size in Inches	12 ft.	14 ft.	16 ft.	18 ft.	20 ft.
1 x 4	4	4⅔	5⅓	6	6⅔
1 x 6	6	7	8	9	10
1 x 8	8	9⅓	10⅔	12	13⅓
1 x 10	10	11⅔	13⅓	15	16⅔
1 x 12	12	14	16	18	20
2 x 3	6	7	8	9	10
2 x 4	8	9⅓	10⅔	12	13⅓
2 x 6	12	14	16	18	20
2 x 8	16	18⅔	21⅓	24	26⅔
2 x 10	20	23⅓	26⅔	30	33⅓
2 x 12	24	28	32	36	40
4 x 4	16	18⅔	21⅓	24	26⅔
6 x 6	36	42	48	54	60
8 x 8	64	74⅔	85⅓	96	106⅔
10 x 10	100	116⅔	133⅓	150	166⅔
12 x 12	144	168	192	216	240